TIME-DOMAIN ELECTROMAGNETIC RECIPROCITY IN ANTENNA MODELING

TIME-DOMAIN ELECTROMAGNETIC RECIPROCITY IN ANTENNA MODELING

MARTIN ŠTUMPF

 IEEE Antennas and Propagation Society, *Sponsor*

The IEEE Press Series on Electromagnetic Wave Theory

IEEE PRESS

WILEY

Published by John Wiley & Sons, Inc., Hoboken, New Jersey.
Published simultaneously in Canada.

For general information on our other products and services or for technical support, please contact our Customer Care Department within the United States at (800) 762-2974, outside the United States at (317) 572-3993 or fax (317) 572-4002.

Wiley also publishes its books in a variety of electronic formats. Some content that appears in print may not be available in electronic formats. For more information about Wiley products, visit our web site at www.wiley.com.

Library of Congress Cataloging-in-Publication Data is available.

hardback: 9781119612315

Printed in the United States of America.

V10013504_082919

Dedicated to my parents

CONTENTS

PREFACE

The present book is meant to give an account of applications of the time-domain electromagnetic reciprocity theorem to solving selected fundamental problems of antenna theory. In order to carefully address the underlying principles of solution methodologies, the complexity of analyzed problems is deliberately kept at fundamental level. It is believed that this strategy that aims at laying sound theoretical foundations of the proposed solutions is vital to their further developments. For the reader's convenience, the text is further supplemented with simple MATLAB® code implementations, thus enabling to explore the beauty of electromagnetic (EM) reciprocity by conducting (numerical) experiments.

The author takes this opportunity to express his thanks to H. A. Lorentz Chair Emeritus Professor Adrianus T. de Hoop, Delft University of Technology, for his stimulating interest in the presented research and to Professor Giulio Antonini, University of L'Aquila, for his kind support. The author also wishes to extend his thanks to Mary Hatcher, Danielle LaCourciere, and Victoria Bradshaw at John Wiley & Sons, Inc., and to Pascal Raj Francois at SPi Global for their professional assistance in the final preparation of this book. This research was funded by the Czech Ministry of Education, Youth and Sports under Grant LO1401. The financial support is gratefully acknowledged.

ACRONYMS

CdH	Cagniard-DeHoop
CdH-MoM	Cagniard-DeHoop Method of Moments
EM	ElectroMagnetic
EMC	ElectroMagnetic Compatibility
FD	Frequency Domain
MoM	Method of Moments
PCB	Printed Circuit Board
PEC	Perfect Electric Conductor
TD	Time Domain
VED	Vertical Electric Dipole

CHAPTER 1

INTRODUCTION

The principle of reciprocity is, in its general sense, interpreted as a norm defining the response to the mutual interaction between two entities involved. Within the realm of electromagnetic (EM) theory, the interacting entities are EM field states, and their mutual interaction is prescribed by Lorentz's reciprocity theorem [1, 2]. The theorem has been recognized as a truly universal relation providing a rigorous basis for approaching both direct and inverse problems of wave field physics [3, 4].

Apart from a limited number of EM field problems that can be solved exactly in terms of analytic functions, the vast majority of EM scattering and radiation problems met in practice must be handled approximately by means of analytical approximations or/and numerical techniques. As the EM reciprocity theorem encompasses all "weak" formulations of the EM (differential) field equations, it offers a convenient venue for developing computational schemes [5]. This strategy has been successfully followed in constructing a general finite-element formulation [6] and dedicated time domain (TD) contour-integral strategies for analyzing planar structures [7], for instance.

A sophisticated analytical method for tackling a large class of problems in wave field physics directly in space-time is known as the Cagniard-DeHoop (CdH) method [8]. It has been shown that the CdH method yields the exact solutions to a large class of canonical TD wave field problems in electromagnetics [9], acoustics [10], and elastodynamics [11]. Since the CdH method is also capable of providing useful large-argument asymptotic solutions [12], as well as insightful closed-form solutions based on the (modified) Kirchhoff approximation [13], a natural question arises whether the CdH approach could also be applicable to constructing a reciprocity-based TD integral-equation technique. Introducing such a novel numerical scheme, hereinafter referred to as the Cagniard-DeHoop method of moments (CdH-MoM), is an essential objective of the present book.

An efficient computational approach for analyzing planarly layered structures in the real frequency domain (FD) is well known as the integral equation technique

Time-Domain Electromagnetic Reciprocity in Antenna Modeling, First Edition. Martin Štumpf.
© 2020 by The Institute of Electrical and Electronics Engineers, Inc. Published 2020 by John Wiley & Sons, Inc.

[14]. Its high computational efficiency stems from the reduction of the solution space to conductive surfaces, which is achieved by including the effect of layering in the pertaining Green's functions constructed traditionally via Sommerfeld's formulation [15, chapter VI]. Although the theory of TD EM field propagation in planarly stratified media is available via the CdH methodology [16], it has never been incorporated into the existing TD integral-equation schemes relying heavily upon the simple form of a Green's function associated with an unbounded, homogeneous, and isotropic medium. Since the CdH-MoM lends itself to the inclusion of mutually parallel layers, it may provide a suitable means to fill the void in computational electromagnetics.

1.1 SYNOPSIS

In the present book, we explore modeling methodologies for analyzing TD EM wave fields associated with fundamental antenna topologies. A common feature of all the methods and solution strategies employed is the use of the EM reciprocity theorem of the time-convolution type as the point of departure.

In chapter 2, the reciprocity theorem is applied to formulate a direct EM scattering problem regarding EM scattering and radiation from a thin-wire antenna. The result is a complex-FD reciprocity relation, the enforcement of the equality in which yields a "weak" solution of the scattering problem. In order to achieve the solution in the (original) TD, we adopt here the strategy behind the CdH method [8]. Namely, the reciprocity relation in the complex-FD is represented in terms of slowness-domain integrals that are subsequently handled analytically with the aid of standard tools of complex analysis. The employed slowness representation is briefly introduced in appendix A, and its illustrative application to the analysis of the electric-current (space-time) distribution along an infinitely-long, gap-excited antenna is given in appendix B. Via the representation, it is shown that a piecewise linear space-time distribution of the induced electric current along a wire antenna can be calculated upon solving a system of equations, whose coefficients are, for thin wire antennas, expressible via elementary functions only. This is demonstrated in appendix C, where the slowness-domain representation of (the elements of) the impedance matrix is transformed back to TD via the CdH method. Chapter 2 is further supplemented with appendix K, where a demo MATLAB® implementation of the introduced solution procedure is provided, including both the antenna excitation via a voltage gap source and a uniform EM plane wave.

The analytical handling of the EM reciprocity relation with the aid of the CdH method is also applied in the subsequent chapter 3, where the pulsed EM coupling between parallel thin-wire antennas is analyzed. This chapter is again supplemented with appendix D, where the elements of the pertaining TD mutual-impedance matrix are described in detail. It is shown that, in contrast to the self-impedance matrix elements described in appendix C, the filling of the mutual-impedance matrix calls, in general, for a numerical calculation of double integrals.

In chapter 4, we propose a straightforward methodology for incorporating ohmic losses of the analyzed thin-wire antenna. It is shown that the effect of losses can be included in by modifying the boundary condition on the cylindrical surface of a wire antenna. This way facilitates the incorporation of losses in a separate impedance matrix, which lends itself to modular programming. This chapter is further supplemented with appendix E, where the internal impedance of a solid EM-penetrable wire is derived with the aid of the gap-excited, thin-wire antenna model and the wave-slowness representation from appendix A.

The EM radiation and scattering characteristic of a wire antenna can be effectively tailored via lumped elements connected to its ports. Therefore, in chapter 5, it is shown how a linear lumped element can be incorporated in the thin-wire CdH-MoM formulation. Again, the presence of a lumped element is captured in an isolated impedance matrix, which makes it possible to evaluate its impact without the need for repeating all the calculations over again. This feature may be profitable especially for optimization routines that require an efficient algorithm for evaluating the objective function.

The numerical procedure introduced in chapter 2 yields the expansion coefficients describing the space-time electric-current distribution along a thin-wire antenna. The thus obtained electric-current distribution may subsequently serve as the input for calculating both the EM radiation and scattering characteristics of the antenna. This is exactly the main objective of chapter 6, where the EM fields radiated from a thin, straight wire segment are expressed in terms of the expansion electric-current coefficients.

In chapter 7, we provide an illustrative numerical example that largely serves for validation of the methodologies described in the previous chapters. Namely, a special form of the (self)-reciprocity antenna relation is first applied to calculate the pulsed EM radiation characteristics from both the antenna self-response, when operating in transmission, and the plane-wave induced response in the relevant receiving situation. Subsequently, the response obtained in the reciprocity-based way is compared with the far-field radiated amplitude computed directly from the electric-current distribution according to chapter 6.

The impact of a wire scatterer on the self-response of a thin-wire transmitting antenna is analyzed in chapter 8. To that end, we make use of a TD reciprocity relation to express the change of the electric-current response of a voltage-gap excited thin-wire antenna using the induced electric-current distribution along the scatterer. The reciprocity-based result is, again, validated directly, via the difference of the electric-current responses calculated in the presence and the absence of the wire scatterer. For the latter, the calculations heavily rely on the methodology introduced in chapter 3.

In chapter 9, we shall analyze the change of EM scattering from a thin-wire antenna due to the change in its localized load. It is first demonstrated that the change of EM scattering characteristics can be expressed using the corresponding EM-radiated field amplitude and the change of the voltage induced across the (varying) antenna lumped load. The result obtained via the reciprocity-based approach is, again, validated directly by evaluating the relevant difference using the methodology specified in section 6.3.

In chapter 10, the EM reciprocity theorem of the time-convolution type is used to evaluate the impact of a wire scatterer on a thin wire receiving antenna. Namely, it is demonstrated that the change of the equivalent Norton electric current can be, in a similar manner as the current response analyzed in chapter 8, related to the electric-current distribution induced along the scatterer. The direct evaluation of (the difference of) the short-circuit electric-current responses validates the TD reciprocity relation.

The following chapter 11 starts by demonstrating that the gap-excited cylindrical antenna (see appendix B) located above the perfectly conducting ground can approximately be handled via transmission-line theory. Under this approximation, the EM reciprocity theorem of the time-convolution type is applied to derive an EM-field-to-line coupling model interrelating the terminal voltage and current quantities with the weighted distribution of the excitation EM wave fields along the transmission line. It is next shown that the (integral) reciprocity-based coupling model can be understood as a generalization of the classic (differential) EM-field-to-line coupling models. Finally, via the EM reciprocity theorem, again, alternative coupling models are introduced, concerning both the EM plane-wave incidence and a prescribed EM volume-source distribution.

In order to provide the reader with straightforward applications of the EM reciprocity–based coupling model, the EM plane-wave–induced Thévenin-voltage response of transmission lines is analyzed in chapter 12. Namely, we derive closed-form expressions for the induced voltages concerning a finite transmission line above the perfect ground and a narrow trace on a grounded dielectric slab. The validity of approximate expressions for the grounded-slab problem configuration is finally discussed with the aid of a three-dimensional computational EM tool.

Whenever the external EM field that couples to a transmission line cannot be longer approximated by a plane wave, sophisticated analytical techniques can be used to evaluate the induced voltage response. This is exactly demonstrated in chapter 13, where the vertical electric dipole (VED)–induced Thévenin's voltage response of a transmission line is analyzed with the help of the CdH method. It is further shown that the effect of a finite ground conductivity can be readily accounted for via the Cooray-Rubinstein formula, thus providing a computationally efficient, analytical model for lightning-induced voltage calculations. The handling of generic integrals necessary for deriving the corresponding TD closed-form expressions is closely described in appendix F. Moreover, the chapter is supplemented with an illustrative numerical example and demo MATLAB$^{®}$ implementations (see appendix L).

In chapter 14, the EM reciprocity theorem is applied to propose a computational technique capable of analyzing planar strip antennas. Following the lines of reasoning similar to those used in chapter 2, it is demonstrated that the electric-current surface density along a narrow perfect electric conductor (PEC) strip follows upon carrying out an updating step-by-step procedure. The elements of the relevant TD "impeditivity matrix" interrelating the induced electric-current surface density with the excitation voltage pulse are closely specified in appendix G with the aid of the CdH method again. This chapter is further supplemented with appendix M, where the reader is provided with an illustrative MATLAB® implementation and appendix H concerning a recursive-convolution method and its numerical implementation. Moreover, it is demonstrated that the proposed CdH-MoM methodology can be readily extended to analyze the performance of a wide-strip antenna supporting a vectorial electric-current surface distribution. Chapter 14 finally concludes with a numerical example demonstrating the validity of the computational procedure via the thin-wire formulation from chapter 2 and the concept of equivalent radius [17].

In case that a strip antenna is not perfectly conducting, the theory of high-contrast, thin-sheet cross-boundary conditions [18] lends itself to incorporate the effect of strip's electric conductivity and permittivity. Therefore, this strategy is adopted in chapter 15, where the impact of a finite conductivity and permittivity is accounted for via an additional impeditivity matrix. Elements of the latter are subsequently derived concerning a homogeneous planar strip with conductive or/and dielectric EM properties and with the Drude-type plasmonic behavior. The relevant thin-sheet jump conditions are justified in appendix I by analyzing a simplified two-dimensional problem configuration.

The inclusion of a linear circuit element in the narrow-strip CdH-MoM formulation is addressed in chapter 16. Pursuing the line of reasoning used in chapter 5, it is demonstrated that a lumped element can be incorporated in the computational scheme by modifying the surface boundary condition at the position where the element is connected. In this way, the impact of a lumped element is taken into account in an isolated impeditivity matrix whose elements are closely specified for a lumped resistor, capacitor, and inductor.

In order to demonstrate that the CdH-MoM is capable of analyzing horizontally layered problem configurations, a narrow strip antenna above a PEC ground plane is analyzed in chapter 17. It is shown that the pertaining impeditivity matrix can be understood as an extension of the one from appendix G that accounts for the effect of reflections from the ground plane. Taking into account the conclusions drawn in section 11.1, the proposed computational scheme is finally validated with the aid of transmission-line theory.

1.2 PREREQUISITES

The mathematical description of EM phenomena is effected via Maxwell field quantities that can be viewed as functions of space and time. To register the position in the analyzed problem configuration, we shall employ a Cartesian reference frame with the origin $\mathcal{O}$ and its (standard) base vectors $\{i_x, i_y, i_z\}$. In line with a common typographic convention, vectors will be hence represented by bold-face italic symbols. Consequently, the position vector, defined as the linear combination of the base vectors, will be represented as $\boldsymbol{x} = x\boldsymbol{i}_x + y\boldsymbol{i}_y + z\boldsymbol{i}_z$, where $\{x, y, z\}$ are the (Cartesian) coordinates specifying the point of observation with respect to the Cartesian frame. The time coordinate is real-valued and is denoted by t. The partial differentiation with respect to a coordinate is denoted by ∂ that is supplied with the pertaining subscript. For example, the spatial differentiation with respect to x is denoted by ∂_x, while the time differentiation will be denoted by ∂_t. The vectorial spatial differentiation operator is then defined as $\boldsymbol{\nabla} = \partial_x\boldsymbol{i}_x + \partial_y\boldsymbol{i}_y + \partial_z\boldsymbol{i}_z$.

A Cartesian vector, or, equivalently, a Cartesian tensor of rank 1, can be arithmetically represented as a 1-D array. For example, $\boldsymbol{v}$ may represent a rank-1 Cartesian tensor (vector) whose components are then $\{v_x, v_y, v_z\}$. A natural extension in this respect is hence a Cartesian tensor of rank 2, also frequently referred to as a dyadic, that is representable as a 2-D array [3, section A.4]. Notationally, such quantities will be further denoted by underlined bold-face italic symbols. For instance, $\underline{\boldsymbol{\eta}}$ will then represent a rank-2 Cartesian tensor. The definition of products of tensors is most conveniently expressed through the subscript notation with the summation convention [3, appendix A]. For a definition of selected products between rank-1 and rank-2 Cartesian-tensor quantities, we refer the reader to Ref. [4, section 1].

1.2.1 One-Sided Laplace Transformation

An alternative way to describe a causal EM space-time quantity is offered by the one-sided Laplace transformation (e.g. [19, section 1.2.1], [3, section B.1]). To introduce the concept, we assume that EM sources that generate the EM wave fields are activated at the origin $t = 0$. Consequently, in view of the property of causality, we will analyze the behavior of the EM space-time quantity, say $f(\boldsymbol{x}, t)$, in $\mathcal{T} = \{t \in \mathbb{R}; t > 0\}$. The one-sided Laplace transformation of the causal quantity is then defined as

$$\hat{f}(\boldsymbol{x}, s) = \int_{t=0}^{\infty} \exp(-st)f(\boldsymbol{x}, t)\mathrm{d}t \tag{1.1}$$

where $\{s \in \mathbb{C}; \mathrm{Re}(s) \geq s_0\}$ denotes the Laplace-transform parameter (complex frequency). A sufficient condition for the integral to exist is its absolute convergence along the line $\{s \in \mathbb{C}; \mathrm{Re}(s) = s_0\}$ parallel to the imaginary axis. In the analysis presented in the book, we shall limit ourselves to (physical) EM wave quantities that are bounded. In other words, we will apply the one-sided Laplace transformation (1.1) only to EM wave quantities of the zero exponential order, that is, $f(\boldsymbol{x}, t) = O(1)$ as $t \to \infty$. Consequently, $\exp(-s_0 t)f(\boldsymbol{x}, t) = o(1)$ as $t \to \infty$ for all values $\{s \in \mathbb{C}; \mathrm{Re}(s) \geq s_0 > 0\}$ of the complex s-plane, to the right of the line of convergence (see Figure 1.1). Hence, in the right half $\{s \in \mathbb{C}; \mathrm{Re}(s) > s_0 > 0\}$ of the complex s-plane, the one-sided Laplace transformation of a causal wave quantity $\hat{f}(\boldsymbol{x}, s)$ does exist and is here, thanks to the analyticity of the transform kernel $\exp(-st)$, an analytic function of s.

Interpreting Eq. (1.1) as an integral equation to be solved for the unknown function $f(\boldsymbol{x}, t)$, a question as to its uniqueness arises. Fortunately, the existence of the

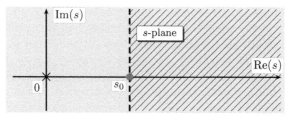

FIGURE 1.1. Complex s-plane with the line of convergence $\mathrm{Re}(s) = s_0$ and the region of regularity $\{s \in \mathbb{C}; \mathrm{Re}(s) > s_0\}$.

one-to-one mapping between a causal quantity $f(\boldsymbol{x}, t)$ and its Laplace transform $\hat{f}(\boldsymbol{x}, s)$ is guaranteed by Lerch's uniqueness theorem provided that the equality in Eq. (1.1) is invoked along the Lerch sequence $\mathcal{L} = \{s \in \mathbb{R}; s = s_0 + nh, h > 0, n = 1, 2, \ldots\}$ [4, appendix]. The choice of taking the Laplace-transform parameter s real-valued and positive is also employed in the CdH method [8] that heavily relies on the uniqueness theorem due to Matyáš Lerch.

With the uniqueness theorem in mind, we shall frequently represent some TD operators in the s-domain. This applies, in particular, to the operation of (continuous) time convolution. The time convolution of two causal wave quantities, say $f(\boldsymbol{x}, t)$ and $g(\boldsymbol{x}, t)$, is defined as

$$[f \underset{t}{*} g](\boldsymbol{x}, t) = \int_{\tau=0}^{t} f(\boldsymbol{x}, \tau) g(\boldsymbol{x}, t - \tau) \mathrm{d}\tau \tag{1.2}$$

for all $t \geq 0$. Applying now (1.1) to Eq. (1.2), we may write

$$\int_{t=0}^{\infty} \exp(-st) \left[\int_{\tau=0}^{t} f(\boldsymbol{x}, \tau) g(\boldsymbol{x}, t - \tau) \mathrm{d}\tau \right] \mathrm{d}t$$

$$= \int_{\tau=0}^{\infty} f(\boldsymbol{x}, \tau) \left[\int_{t=0}^{\infty} \exp(-st) g(\boldsymbol{x}, t - \tau) \mathrm{d}t \right] \mathrm{d}\tau$$

$$= \int_{\tau=0}^{\infty} \exp(-s\tau) f(\boldsymbol{x}, \tau) \mathrm{d}\tau \int_{\vartheta=0}^{\infty} \exp(-s\vartheta) g(\boldsymbol{x}, \vartheta) \mathrm{d}\vartheta$$

$$= \hat{f}(\boldsymbol{x}, s) \hat{g}(\boldsymbol{x}, s) \tag{1.3}$$

where we have first interchanged the order of the integrations with respect to t and τ, which is permissible thanks to the absolute convergence of the Laplace transforms of both $f(\boldsymbol{x}, t)$ and $g(\boldsymbol{x}, t)$, and, secondly, we have used the substitution $\vartheta = t - \tau$. For the Laplace transforms in Eq. (1.3) to make any sense, there clearly must be at least one value of s at which $\hat{f}(\boldsymbol{x}, s)$ and $\hat{g}(\boldsymbol{x}, s)$ do converge simultaneously. If $f = O[\exp(\alpha t)]$ as $t \to \infty$ for some $\alpha \in \mathbb{R}$ and $g = O[\exp(\beta t)]$ as $t \to \infty$ for some $\beta \in \mathbb{R}$, the region of convergence for Eq. (1.3) extends over the half-plane to the right of the line of convergence $\mathrm{Re}(s) = \max\{\alpha, \beta\}$. Again, if both $f(\boldsymbol{x}, t)$ and

$g(\boldsymbol{x}, t)$ are bounded functions whose exponential orders are $\alpha = \beta = 0$, then their region of regularity extends over the right half of the complex s-plane. The property that the Laplace transform of the time convolution of two causal wave quantities is equal to the product of their Laplace transforms (see Eq. (1.3)) will be frequently (and tacitly) used throughout the book. For further useful properties of the Laplace transform, we refer the reader to more complete accounts on this subject (e.g. [3, appendix B.1], [20]).

1.2.2 Lorentz's Reciprocity Theorem

The point of departure for analyzing the space-time EM problems posed in the present book is the TD reciprocity theorem of the time-convolution type (see [3, section 28.2] [4, section 1.4.1]) that is in literature widely known as Lorentz's reciprocity theorem (e.g. [21, section 8.6]).

To introduce the relation, let us assume two states of EM fields, say (A) and (B), that are on a spatial domain $\mathcal{D}$ governed by Maxwell's EM field equations in the s-domain [3, section 24.4]

$$-\boldsymbol{\nabla} \times \hat{\boldsymbol{H}}^{\mathrm{A,B}} + \hat{\underline{\eta}}^{\mathrm{A,B}} \cdot \hat{\boldsymbol{E}}^{\mathrm{A,B}} = -\hat{\boldsymbol{J}}^{\mathrm{A,B}} \tag{1.4}$$

$$\boldsymbol{\nabla} \times \hat{\boldsymbol{E}}^{\mathrm{A,B}} + \hat{\underline{\zeta}}^{\mathrm{A,B}} \cdot \hat{\boldsymbol{H}}^{\mathrm{A,B}} = -\hat{\boldsymbol{K}}^{\mathrm{A,B}} \tag{1.5}$$

for all $\boldsymbol{x} \in \mathcal{D}$, in which

- $\hat{\boldsymbol{E}}^{\mathrm{A,B}}$ = electric field strength in V/m;
- $\hat{\boldsymbol{H}}^{\mathrm{A,B}}$ = magnetic field strength in A/m;
- $\hat{\boldsymbol{J}}^{\mathrm{A,B}}$ = electric current volume density in A/m^2;
- $\hat{\boldsymbol{K}}^{\mathrm{A,B}}$ = magnetic current volume density in V/m^2;
- $\hat{\underline{\eta}}^{\mathrm{A,B}}$ = transverse admittance (per unit length) of the medium in S/m;
- $\hat{\underline{\zeta}}^{\mathrm{A,B}}$ = longitudinal impedance (per unit length) of the medium in Ω/m.

The EM field Eqs. (1.4) and (1.5) are further supplemented with the constitutive relations

$$\hat{\underline{\eta}}^{\mathrm{A,B}}(\boldsymbol{x}, s) = \hat{\underline{\sigma}}^{\mathrm{A,B}}(\boldsymbol{x}, s) + s\hat{\underline{\epsilon}}^{\mathrm{A,B}}(\boldsymbol{x}, s) \tag{1.6}$$

$$\hat{\underline{\zeta}}^{\mathrm{A,B}}(\boldsymbol{x}, s) = s\hat{\underline{\mu}}^{\mathrm{A,B}}(\boldsymbol{x}, s) \tag{1.7}$$

for all $\boldsymbol{x} \in \mathcal{D}$, in which

- $\hat{\underline{\sigma}}^{\mathrm{A,B}}$ = electric conductivity in S/m;
- $\hat{\underline{\epsilon}}^{\mathrm{A,B}}$ = electric permittivity in F/m;
- $\hat{\underline{\mu}}^{\mathrm{A,B}}$ = permeability in H/m.

The point of departure is the following local interaction quantity [3, section 28.4]

$$\boldsymbol{\nabla} \cdot \left[\hat{\boldsymbol{E}}^{\mathrm{A}}(\boldsymbol{x}, s) \times \hat{\boldsymbol{H}}^{\mathrm{B}}(\boldsymbol{x}, s) - \hat{\boldsymbol{E}}^{\mathrm{B}}(\boldsymbol{x}, s) \times \hat{\boldsymbol{H}}^{\mathrm{A}}(\boldsymbol{x}, s) \right] \tag{1.8}$$

applying throughout $\mathcal{D}$ that can be with the aid of Eqs. (1.4)–(1.5) written as

$$\begin{aligned}
\boldsymbol{\nabla} \cdot &\left[\hat{\boldsymbol{E}}^{\mathrm{A}}(\boldsymbol{x}, s) \times \hat{\boldsymbol{H}}^{\mathrm{B}}(\boldsymbol{x}, s) - \hat{\boldsymbol{E}}^{\mathrm{B}}(\boldsymbol{x}, s) \times \hat{\boldsymbol{H}}^{\mathrm{A}}(\boldsymbol{x}, s) \right] \\
&= \hat{\boldsymbol{H}}^{\mathrm{A}}(\boldsymbol{x}, s) \cdot \left\{ \hat{\underline{\zeta}}^{\mathrm{B}}(\boldsymbol{x}, s) - [\hat{\underline{\zeta}}^{\mathrm{A}}(\boldsymbol{x}, s)]^{\mathcal{T}} \right\} \cdot \hat{\boldsymbol{H}}^{\mathrm{B}}(\boldsymbol{x}, s) \\
&\quad - \hat{\boldsymbol{E}}^{\mathrm{A}}(\boldsymbol{x}, s) \cdot \left\{ \hat{\underline{\eta}}^{\mathrm{B}}(\boldsymbol{x}, s) - [\hat{\underline{\eta}}^{\mathrm{A}}(\boldsymbol{x}, s)]^{\mathcal{T}} \right\} \cdot \hat{\boldsymbol{E}}^{\mathrm{B}}(\boldsymbol{x}, s) \\
&\quad + \hat{\boldsymbol{J}}^{\mathrm{A}}(\boldsymbol{x}, s) \cdot \hat{\boldsymbol{E}}^{\mathrm{B}}(\boldsymbol{x}, s) - \hat{\boldsymbol{K}}^{\mathrm{A}}(\boldsymbol{x}, s) \cdot \hat{\boldsymbol{H}}^{\mathrm{B}}(\boldsymbol{x}, s) \\
&\quad - \hat{\boldsymbol{J}}^{\mathrm{B}}(\boldsymbol{x}, s) \cdot \hat{\boldsymbol{E}}^{\mathrm{A}}(\boldsymbol{x}, s) + \hat{\boldsymbol{K}}^{\mathrm{B}}(\boldsymbol{x}, s) \cdot \hat{\boldsymbol{H}}^{\mathrm{A}}(\boldsymbol{x}, s)
\end{aligned} \tag{1.9}$$

for all $\boldsymbol{x} \in \mathcal{D}$ and $^{\mathcal{T}}$ is the transpose operator. The global form of the interaction quantity is derived upon integrating the local interaction over the union of sub domains constituting $\mathcal{D}$ in each of which we assume that the terms of Eq. (1.9) are continuous functions with respect to $\boldsymbol{x}$. Upon applying Gauss' divergence theorem and adding the contributions of the integrations, we arrive at

$$\begin{aligned}
\int_{\boldsymbol{x} \in \partial \mathcal{D}} &\left(\hat{\boldsymbol{E}}^{\mathrm{A}} \times \hat{\boldsymbol{H}}^{\mathrm{B}} - \hat{\boldsymbol{E}}^{\mathrm{B}} \times \hat{\boldsymbol{H}}^{\mathrm{A}} \right) \cdot \boldsymbol{\nu} \, \mathrm{d}A \\
&= \int_{\boldsymbol{x} \in \mathcal{D}} \left\{ \hat{\boldsymbol{H}}^{\mathrm{A}} \cdot \left[\hat{\underline{\zeta}}^{\mathrm{B}} - (\hat{\underline{\zeta}}^{\mathrm{A}})^{\mathcal{T}} \right] \cdot \hat{\boldsymbol{H}}^{\mathrm{B}} \right. \\
&\qquad \left. - \hat{\boldsymbol{E}}^{\mathrm{A}} \cdot \left[\hat{\underline{\eta}}^{\mathrm{B}} - (\hat{\underline{\eta}}^{\mathrm{A}})^{\mathcal{T}} \right] \cdot \hat{\boldsymbol{E}}^{\mathrm{B}} \right\} \mathrm{d}V \\
&\quad + \int_{\boldsymbol{x} \in \mathcal{D}} \left(\hat{\boldsymbol{J}}^{\mathrm{A}} \cdot \hat{\boldsymbol{E}}^{\mathrm{B}} - \hat{\boldsymbol{K}}^{\mathrm{A}} \cdot \hat{\boldsymbol{H}}^{\mathrm{B}} \right. \\
&\qquad \left. - \hat{\boldsymbol{J}}^{\mathrm{B}} \cdot \hat{\boldsymbol{E}}^{\mathrm{A}} + \hat{\boldsymbol{K}}^{\mathrm{B}} \cdot \hat{\boldsymbol{H}}^{\mathrm{A}} \right) \mathrm{d}V
\end{aligned} \tag{1.10}$$

where we have invoked the continuity of $\boldsymbol{\nu} \times \hat{\boldsymbol{E}}^{\mathrm{A,B}}$ and $\boldsymbol{\nu} \times \hat{\boldsymbol{H}}^{\mathrm{A,B}}$ across the common interfaces of the subdomains. The resulting relation (1.10) will be further referred to as the EM reciprocity theorem of the time convolution type or Lorentz's theorem in short. Its left-hand side consists of contributions from the outer boundary of domain $\mathcal{D}$ that is denoted by $\partial \mathcal{D}$. The first integral on the right-hand side then represents the contrasts in the EM properties of the media in states (A) and (B). For the sake of conciseness, this term will be referred to as the interaction of the field and material states. Apparently, the field-material

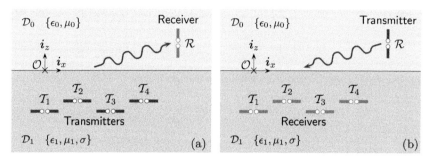

FIGURE 1.2. An example of the use of EM reciprocity to replace (a) the actual problem with an (b) equivalent one that is easier to analyze.

interaction vanishes whenever $\hat{\underline{\zeta}}^{\mathrm{B}} = (\hat{\underline{\zeta}}^{\mathrm{A}})^{\mathcal{T}}$ and $\hat{\underline{\eta}}^{\mathrm{B}} = (\hat{\underline{\eta}}^{\mathrm{A}})^{\mathcal{T}}$ for all $x \in \mathcal{D}$, that is, whenever the media in the both states are each other's adjoint. Finally, the second integral on the right-hand side of Eq. (1.10) is interpreted as the interaction of the field and source states.

Having defined the interacting EM field states and the domain to which the reciprocity theorem applies, Eq. (1.10) can be established. Depending on the choice of EM field states, the resulting relation can be interpreted as a mere relation, an integral representation, an integral equation, or, eventually, a complete solution. Actually, a wide range of venues offered by reciprocity relations is exactly the reason why they are among the most intriguing relations in wave field physics. Since the use of EM reciprocity is largely a matter of ingenuity, it is hardly possible to give a comprehensive application manual. One may, however, provide some typical choices of (A) and (B) states covering a broad spectrum of applications. In computational electromagnetics, for instance, state (A) is typically associated with the (actual) scattered EM wave fields via induced (unknown) current densities, while state (B) is taken to be the computational (or testing) state representing the manner in which the EM field quantities in state (A) are calculated [5]. This strategy is also followed in the present book to formulate a CdH-method–based TD integral equation technique. In antenna theory, states (A) and (B) typically represent receiving and transmitting modes of an antenna system [4]. Similarly, in the context of electromagnetic compatibility (EMC), the reciprocity theorem may serve to link susceptibility and emission properties of the analyzed system at hand. A result from this category is the EM-field-to-line coupling model introduced in chapter 11.

The reciprocity theorem is frequently exploited to replace the actual (tough) task by an equivalent one that is smoothly amenable to an analytical analysis, to measurements or computer simulations. A typical problem from this category is the EM coupling between (a set of) insulated transmitting antennas and a receiving (victim) antenna above the conductive layer (see Figure 1.2), which is of interest to designing wireless inter chip or submarine communication systems [22–24]. In accordance with linear time-invariant system theory, the interaction between such antennas can be characterized in terms of transfer-impedance matrices that are, in virtue of EM reciprocity, symmetrical [4, chapter 7]. Accordingly, instead of carrying out multiple analyses to evaluate the EM field transfers from each buried antenna to the receiver, it may be more convenient to analyze the equivalent problem (see Figure 1.2b) in a single simulation. Furthermore, the equivalent problem configuration, where the antenna above the interface acts as a transmitter, is suitable for its approximate analysis. Indeed, if the medium in $\mathcal{D}_1$ described by its (scalar, real-valued, and positive) permittivity ϵ_1, conductivity σ, and permeability μ_0 is sufficiently (electromagnetically) dense with respect to the one in the upper half-space $\mathcal{D}_0$, then the EM field penetrated in $\mathcal{D}_1$ varies dominantly in the normal direction with respect to the interface, thus resembling a plane wave. Consequently, the relevant electric-field strength (i.e. polarized along the insulated antennas' axes) as observed at a horizontal offset $r > 0$ and at a depth $z = -\zeta < 0$ can approximately be expressed via the electric-field distribution at the level of the interface, that is

$$E_x(r, -\zeta, t) \simeq E_x(r, 0, t) *_t \Psi(\zeta, t) \tag{1.11}$$

where (cf. [3, eq. (26.5–29)])

$$\Psi(\zeta, t) = \delta(t - \zeta/c_1) \exp(-\alpha t/2)$$
$$+ \frac{\alpha\zeta/2c_1}{(t^2 - \zeta^2/c_1^2)^{1/2}} I_1[\alpha(t^2 - \zeta^2/c_1^2)^{1/2}/2] \exp(-\alpha t/2) H(t - \zeta/c_1) \tag{1.12}$$

with $I_1(t)$ being the modified Bessel function of the first kind and order one, $H(t)$ denotes the Heaviside unit-step function, $\alpha = \sigma/\epsilon_1$ and $c_1 = (\mu_0\epsilon_1)^{-1/2}$. Furthermore, under the assumption given previously, the electric-field distribution at $z = 0$ can be related to the corresponding magnetic-field strength via the surface-impedance (Leontovich) boundary condition

$$E_x(r, 0, t) \simeq -Z(t) *_t H_y(r, 0, t) \tag{1.13}$$

where $Z(t) = \zeta_1 \partial_t[\mathsf{I}_0(\alpha t/2) H(t)]$ with $\zeta_1 = (\mu_0/\epsilon_1)^{1/2}$ is the TD surface impedance described via $\mathsf{I}_0(t) \triangleq I_0(t) \exp(-t)$, which is defined as the (scaled) modified Bessel function of the first kind and order zero (see [25, figure 9.8]). Upon combining Eq. (1.11) with the surface-impedance boundary condition (1.13), the horizontal component of the electric-field strength in the lossy medium can be related to the magnetic-field distribution over the planar interface, that is

$$E_x(r, -\zeta, t) \simeq -Z(t) *_t H_y(r, 0, t) *_t \Psi(\zeta, t) \tag{1.14}$$

Under certain conditions that are met for typical lightning-induced calculations [26, 27], for instance, the resulting approximate expression (1.14) can be further simplified by replacing the actual magnetic-field distribution at $z = 0$ with the one

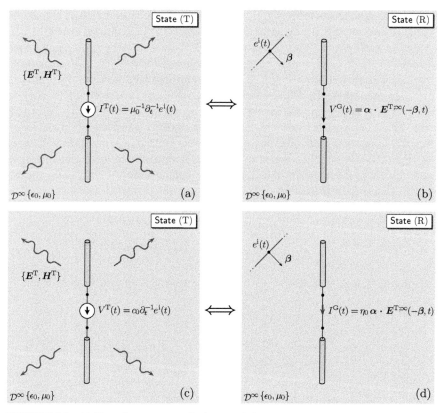

FIGURE 1.3. (a) Electric-current–excited transmitting antenna and (b) the corresponding receiving antenna whose open-circuit voltage response is related to the radiation characteristics; (c) voltage-excited transmitting antenna and (d) the corresponding receiving antenna whose short-circuit electric-current response is related to the radiation characteristics.

pertaining to the PEC surface. As the latter field distribution can be for many fundamental EM sources expressed in closed form, a compact approximate expression for $E_x(r, -\zeta, t)$ follows.

Analyzing mutually reciprocal scenarios is not only useful for simplifying the problem solution itself but may also be beneficial for the validation of purely computational tools. Typical transmitting (T) and receiving (R) scenarios regarding a wire antenna that can be analyzed for validation purposes are shown in Figure 1.3. In them, the transmitting antenna radiating EM wave fields $\{\boldsymbol{E}^{\mathrm{T}}, \boldsymbol{H}^{\mathrm{T}}\}$, whose far-field amplitudes are denoted by $\{\boldsymbol{E}^{\mathrm{T};\infty}, \boldsymbol{H}^{\mathrm{T};\infty}\}$ (see [3, section 26.12]), are related to port responses induced by a uniform EM plane wave defined by its polarization vector $\boldsymbol{\alpha}$, by a unit vector in the direction of propagation $\boldsymbol{\beta}$ and its pulse shape $e^{\mathrm{i}}(t)$. Consequently, by virtue of EM reciprocity, the open-circuit (Thévenin) plane-wave

TABLE 1.1. Application of the Reciprocity Theorem

	Domain $\mathcal{D}$	
Time-Convolution	State (A)	State (B)
Source	$\{\hat{\boldsymbol{J}}^{\mathrm{A}}, \hat{\boldsymbol{K}}^{\mathrm{A}}\}$	$\{\hat{\boldsymbol{J}}^{\mathrm{B}}, \hat{\boldsymbol{K}}^{\mathrm{B}}\}$
Field	$\{\hat{\boldsymbol{E}}^{\mathrm{A}}, \hat{\boldsymbol{H}}^{\mathrm{A}}\}$	$\{\hat{\boldsymbol{E}}^{\mathrm{B}}, \hat{\boldsymbol{H}}^{\mathrm{B}}\}$
Material	$\{\hat{\underline{\boldsymbol{\eta}}}^{\mathrm{A}}, \hat{\underline{\boldsymbol{\zeta}}}^{\mathrm{A}}\}$	$\{\hat{\underline{\boldsymbol{\eta}}}^{\mathrm{B}}, \hat{\underline{\boldsymbol{\zeta}}}^{\mathrm{B}}\}$

induced voltage response pertaining to the receiving state can be directly related to the co-polarized far-field amplitude observed in the backward direction

$$V^{\mathrm{G}}(t) = \boldsymbol{\alpha} \cdot \boldsymbol{E}^{\mathrm{T};\infty}(-\boldsymbol{\beta}, t) \tag{1.15}$$

provided that the exciting electric-current pulse $I^{\mathrm{T}}(t)$ is proportional to the (time-integrated) plane-wave signature according to

$$I^{\mathrm{T}}(t) = \mu_0^{-1} \partial_t^{-1} e^{\mathrm{i}}(t) \tag{1.16}$$

where ∂_t^{-1} denotes the time-integration operator [28, eq. (20)]. The corresponding scenarios are depicted in Figure 1.3a and b. Alternatively, if the port of the receiving wire antenna is short-circuited, the induced (Norton) electric-current response can be found from

$$I^{\mathrm{G}}(t) = \eta_0 \, \boldsymbol{\alpha} \cdot \boldsymbol{E}^{\mathrm{T};\infty}(-\boldsymbol{\beta}, t) \tag{1.17}$$

with $\eta_0 = (\epsilon_0/\mu_0)^{1/2}$ provided that the excitation voltage pulse $V^{\mathrm{T}}(t)$ is related to the plane-wave pulse shape via (see Figure 1.3c and d)

$$V^{\mathrm{T}}(t) = c_0 \partial_t^{-1} e^{\mathrm{i}}(t) \tag{1.18}$$

Yet another and more general example interrelating the transmitting (T) and receiving (R) states of a wire antenna is numerically analyzed in chapter 7. In conclusion, the EM reciprocity theorem may also be viewed as a powerful tool for validation of EM solvers. This (reciprocity-based) strategy has been applied in chapters 7, 8, 9, and 10 to check the consistency of the proposed solution methodologies.

To ease the use of EM reciprocity, we will adopt the tabular representation of the EM reciprocity theorem (see [4, 19]). In this fashion, the source, field, and material states pertaining to the interacting EM field states are clearly summarized in a table. For instance, the table corresponding to the generic form of the EM reciprocity theorem of the time-convolution type (see Eq. (1.10)) is given here as Table 1.1.

CHAPTER 2

CAGNIARD-DEHOOP METHOD OF MOMENTS FOR THIN-WIRE ANTENNAS

The main purpose of this chapter is to develop a reciprocity-based TD integral equation technique for analyzing pulsed EM scattering from a straight thin-wire segment. To that end, we employ the reciprocity theorem of the time-convolution type (see [3, section 28.4] and [4, section 1.4.1]) and formulate the antenna problem via the interaction between the (actual) scattered field and the (computational) testing field. The resulting interaction quantity is subsequently represented using a wave-slowness representation taken along the wire's axis. It is next shown that for appropriate expansion functions, the transform-domain interaction quantity can be evaluated analytically in closed form via the CdH technique [8]. In this way, we will end up with a system of equations whose solution is attainable in an updating, step-by-step manner. The resulting computational procedure will further be referred to as the Cagniard-DeHoop Method of Moments (CdH-MoM).

2.1 PROBLEM DESCRIPTION

The problem under consideration is shown in Figure 2.1. Owing to the rotational symmetry of the antenna structure, the position in the problem configuration will be specified by the radial distance $r = (x^2 + y^2)^{1/2} > 0$ and the axial coordinate z. The antenna structure extends along the z-axis from $z = -\ell/2$ to $z = \ell/2$, where $\ell > 0$ denotes its length. The wire antenna has a circular cross-section of radius $a > 0$. The antenna is embedded in an unbounded, homogeneous, loss-less, and isotropic embedding, denoted by $\mathcal{D}^\infty$, whose EM properties are described by (real-valued, positive and scalar) electric permittivity ϵ_0 and magnetic

Time-Domain Electromagnetic Reciprocity in Antenna Modeling, First Edition. Martin Štumpf.
© 2020 by The Institute of Electrical and Electronics Engineers, Inc. Published 2020 by John Wiley & Sons, Inc.

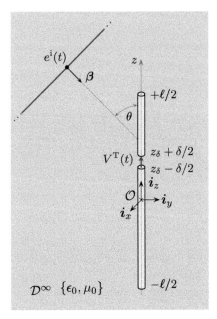

FIGURE 2.1. A straight wire antenna excited by an impulsive plane wave and a voltage gap source.

permeability μ_0. The corresponding EM wave speed is $c_0 = (\epsilon_0\mu_0)^{-1/2} > 0$. The antenna structure is from its exterior domain separated by a closed surface $\mathcal{S}_0$.

In a standard way, we next introduce the scattered EM wave fields, $\{\hat{\boldsymbol{E}}^{\mathrm{s}}, \hat{\boldsymbol{H}}^{\mathrm{s}}\}$, as the difference between the total EM wave fields in the configuration and the incident EM wave fields, denoted by $\{\hat{\boldsymbol{E}}^{\mathrm{i}}, \hat{\boldsymbol{H}}^{\mathrm{i}}\}$. The incident EM wave field that excites the wire antenna can be represented by a uniform EM plane wave or/and by a localized voltage source in the narrow gap whose center is at $z = z_\delta$ with $\{-\ell/2 < z_\delta < \ell/2\}$.

2.2   PROBLEM FORMULATION

The problem of EM scattering from the wire antenna will be analyzed with the aid of the reciprocity theorem of the time-convolution type [4, section 1.4.1]. Therefore, we apply the theorem to the unbounded domain exterior to the antenna and to the scattered field state (s) and the testing state (B) (see Table 2.1). Since the contribution from the exterior "sphere at infinity" vanishes [4, section 1.4.3], we end up with

$$\int_{\boldsymbol{x}\in\mathcal{S}_0} (\hat{\boldsymbol{E}}^{\mathrm{s}} \times \hat{\boldsymbol{H}}^{\mathrm{B}} - \hat{\boldsymbol{E}}^{\mathrm{B}} \times \hat{\boldsymbol{H}}^{\mathrm{s}}) \cdot \boldsymbol{\nu}\mathrm{d}A = 0 \tag{2.1}$$

TABLE 2.1. Application of the Reciprocity Theorem

Domain Exterior to $\mathcal{S}_0$		
Time-Convolution	State (s)	State (B)
Source	0	0
Field	$\{\hat{\boldsymbol{E}}^{\mathrm{s}}, \hat{\boldsymbol{H}}^{\mathrm{s}}\}$	$\{\hat{\boldsymbol{E}}^{\mathrm{B}}, \hat{\boldsymbol{H}}^{\mathrm{B}}\}$
Material	$\{\epsilon_0, \mu_0\}$	$\{\epsilon_0, \mu_0\}$

The interaction quantity can further be rewritten in terms of equivalent electric-current densities, that is, $\partial \hat{\boldsymbol{J}}^{\mathrm{s,B}}(\boldsymbol{x}, s) \triangleq \boldsymbol{\nu}(\boldsymbol{x}) \times \hat{\boldsymbol{H}}^{\mathrm{s,B}}(\boldsymbol{x}, s)$ with $\boldsymbol{x}$ approaching the antenna surface $\mathcal{S}_0$ from the exterior domain $\mathcal{D}^\infty$, as

$$\int_{\boldsymbol{x} \in \mathcal{S}_0} (\hat{\boldsymbol{E}}^{\mathrm{B}} \cdot \partial \hat{\boldsymbol{J}}^{\mathrm{s}} - \hat{\boldsymbol{E}}^{\mathrm{s}} \cdot \partial \hat{\boldsymbol{J}}^{\mathrm{B}}) \, \mathrm{d}A = 0 \tag{2.2}$$

If the radius of the wire is relatively small, we may neglect the electric currents at the end faces of the wire and employ the "reduced form" of (one-dimensional) Pocklington's integral equation [29]. Following these lines of reasoning, the interaction quantity is further approximated by

$$\int_{z=-\ell/2}^{\ell/2} \left[\hat{E}_z^{\mathrm{B}}(a, z, s) \hat{I}^{\mathrm{s}}(z, s) - \hat{E}_z^{\mathrm{s}}(a, z, s) \hat{I}^{\mathrm{B}}(z, s) \right] \mathrm{d}z = 0 \tag{2.3}$$

where $\hat{I}^{\mathrm{s}}(z, s)$ is the (unknown) induced electric current along the antenna axis, and the testing electric-field strength $\hat{E}_z^{\mathrm{B}}(r, z, s)$ is related to the testing current, $\hat{I}^{\mathrm{B}}(z, s)$, according to [3, section 26.9]

$$\hat{E}_z^{\mathrm{B}}(r, z, s) = -s\mu_0 \hat{G}(r, z, s) *_z \hat{I}^{\mathrm{B}}(z, s)$$
$$+ (s\epsilon_0)^{-1} \partial_z^2 \, \hat{G}(r, z, s) *_z \hat{I}^{\mathrm{B}}(z, s) \tag{2.4}$$

where $*_z$ denotes the spatial convolution along the axial z-direction, and the support of the testing current extends over the finite interval $\{-\ell/2 < z < \ell/2\}$. Finally, the (reduced) kernel has the form

$$\hat{G}(r, z, s) = \exp[-s(r^2 + z^2)^{1/2}/c_0]/[4\pi(r^2 + z^2)^{1/2}] \tag{2.5}$$

with s being the real-valued and positive Laplace-transform parameter.

The reciprocity relation (2.3) with Eqs. (2.4) and (2.5) will further be expressed in the so-called slowness domain. To that end, we represent the electric-field strengths in Eq. (2.3) via (A.1) and change the order of integrations in the resulting expression. This way leads to the following reciprocity relation

$$
\frac{s}{2i\pi} \int_{p=-i\infty}^{i\infty} \tilde{E}_z^{\mathrm{B}}(a, p, s) \tilde{I}^{\mathrm{s}}(-p, s) \mathrm{d}p
$$

$$
= \frac{s}{2i\pi} \int_{p=-i\infty}^{i\infty} \tilde{E}_z^{\mathrm{s}}(a, p, s) \tilde{I}^{\mathrm{B}}(-p, s) \mathrm{d}p \tag{2.6}
$$

along with the transform-domain counterpart of Eq. (2.4), that is

$$
\tilde{E}_z^{\mathrm{B}}(r, p, s) = -s\mu_0 \tilde{G}(r, p, s) \tilde{I}^{\mathrm{B}}(p, s)
$$
$$
+ (s^2 p^2 / s\epsilon_0) \tilde{G}(r, p, s) \tilde{I}^{\mathrm{B}}(p, s) \tag{2.7}
$$

where $\tilde{G}(r, p, s)$ is given by Eq. (A.3). The transform-domain reciprocity relation (2.6) is the point of departure for the CdH-MoM described in the ensuing section.

2.3   PROBLEM SOLUTION

In order to solve the problem numerically, the solution domain extending along the wire axis is discretized into $N+1$ segments of a constant length $\Delta = \ell/(N+1) > 0$ (see Figure 2.2a). The points along the uniform grid can be specified by

$$
z_n = -\ell/2 + n \Delta \quad \text{for } n = \{0, 1, \ldots, N+1\} \tag{2.8}
$$

which for $n = 0$ and $n = N+1$ describes the end points, where the end conditions apply, that is

$$
\hat{I}^{\mathrm{s,B}}(\pm \ell/2, s) = 0 \tag{2.9}
$$

Accordingly, the induced electric current is in space expanded in piecewise linear basis functions $\Lambda^{[n]}(z)$ defined as

$$
\Lambda^{[n]}(z) = \begin{cases} 1 + (z - z_n)/\Delta & \text{for } z \in [z_{n-1}, z_n] \\ 1 - (z - z_n)/\Delta & \text{for } z \in [z_n, z_{n+1}] \end{cases} \tag{2.10}
$$

along the inner discretization points for $n = \{1, \ldots, N\}$ (see Figure 2.2b). Likewise, the time axis $\{t \in \mathbb{R}; t > 0\}$ is discretized uniformly with the constant time step $\{t_k = k\Delta t; \Delta t > 0, k = 1, 2, \ldots, M\}$, and the temporal behavior of the unknown current is approximated through a set of triangular functions with

$$
\Lambda_k(t) = \begin{cases} 1 + (t - t_k)/\Delta t & \text{for } t \in [t_{k-1}, t_k] \\ 1 - (t - t_k)/\Delta t & \text{for } t \in [t_k, t_{k+1}] \end{cases} \tag{2.11}
$$

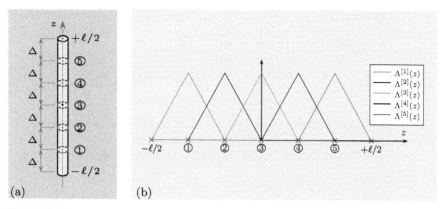

FIGURE 2.2. (a) Uniform discretization of the spatial solution domain; (b) piecewise-linear basis functions.

Finally, the induced electric current is expanded in piecewise linear functions both in space and time, which in the transform domain is described by

$$\tilde{I}^{s}(p,s) \simeq \sum_{n=1}^{N}\sum_{k=1}^{M} i_{k}^{[n]}\tilde{\Lambda}^{[n]}(p)\hat{\Lambda}_{k}(s) \tag{2.12}$$

where $i_{k}^{[n]}$ are unknown coefficients (in A) to be determined.

Furthermore, the testing electric current is chosen to have the piecewise linear spatial distribution and the Dirac-delta behavior in time. Consequently, the testing electric current in the transform domain reads

$$\tilde{I}^{B}(p,s) = \tilde{\Lambda}^{[S]}(p) \tag{2.13}$$

for all $S = \{1, \ldots, N\}$. Substitution of Eqs. (2.12) and (2.13) in the transform-domain reciprocity relation (2.6) leads to the system of complex FD equations, constituents of which can be transformed to the TD with the aid of the CdH technique. In this way, we arrive at

$$\sum_{k=1}^{m}(\underline{Z}_{m-k+1} - 2\underline{Z}_{m-k} + \underline{Z}_{m-k-1}) \cdot \boldsymbol{I}_{k} = \boldsymbol{V}_{m} \tag{2.14}$$

where $\underline{Z}_{k}$ denotes a 2-D $[N \times N]$ impedance array at $t = t_{k}$, $\boldsymbol{I}_{k}$ is an 1-D $[N \times 1]$ array of the current coefficients at $t = t_{k}$, and, finally, $\boldsymbol{V}_{m}$ is an 1-D $[N \times 1]$ array describing the antenna excitation at $t = t_{m}$. The elements of the excitation array $\boldsymbol{V}_{m}$ will be specified later for both the incident EM plane wave and the delta gap source. Equation (2.14) can be solved, by virtue of causality, in an iterative manner for all $m = \{1, \dots, M\}$. Along these lines, we get

$$I_m = \underline{Z}_1^{-1} \cdot \left[V_m - \sum_{k=1}^{m-1} (\underline{Z}_{m-k+1} - 2\underline{Z}_{m-k} + \underline{Z}_{m-k-1}) \cdot I_k \right] \qquad (2.15)$$

for all $m = \{1, \dots, M\}$, from which the actual vector of the electric-current coefficients follows upon inverting the impedance matrix evaluated at $t = t_1 = \Delta t$. The elements of the TD impedance matrix are closely described in appendix C.

2.4 ANTENNA EXCITATION

In this section, we shall analyze the excitation of the wire antenna by a uniform EM plane wave and by a voltage delta gap source (see Figure 2.1). In the analysis, whose objective is to provide TD expressions specifying the elements of the excitation vector V_m (see Eq. (2.15)), we shall evaluate the right-hand side of Eq. (2.6), where we substitute the slowness-domain counterpart of the explicit-type boundary condition

$$\lim_{r \downarrow a} \hat{E}_z^s(r, z, s) = -\lim_{r \downarrow a} \hat{E}_z^i(r, z, s) \qquad (2.16)$$

applying along the PEC surface of the wire antenna for $\{-\ell/2 \le z \le \ell/2\}$. A more general boundary condition addressing EM scattering from an EM-penetrable wire structure will be studied in chapter 4. The two types of excitation will be next discussed separately.

2.4.1 Plane-Wave Excitation

Owing to the rotational symmetry of the problem configuration, the incident EM field distribution along the wire axis can be described as

$$\hat{E}_z^i(0, z, s) = \hat{e}^i(s) \sin(\theta) \exp[sp_0(z - \ell/2)] \qquad (2.17)$$

where we have assumed that the plane wave hits the antenna's top end at instant $t = 0$ and $p_0 = \cos(\theta)/c_0$. Through the residue theorem [30, section 3.11], it is then straightforward to verify that the transform-domain counterpart of Eq. (2.17) reads

$$\tilde{E}_z^i(0, p, s) = \frac{\hat{e}^i(s)}{s} \frac{\exp(sp\ell/2) - \exp(-sp\ell/2) \exp(-sp_0\ell)}{p + p_0} \sin(\theta) \qquad (2.18)$$

The latter is, via the explicit-type boundary condition, subsequently used to evaluate the right-hand side of Eq. (2.6) for the plane-wave incidence. Making use of the testing current distribution represented by Eq. (2.13) and assuming the contour indentation shown in Figure C.1, Cauchy's formula [30, section 2.41] yields

$$\frac{s}{2i\pi} \int_{p=-i\infty}^{i\infty} \tilde{E}_z^s(0,p,s)\tilde{I}^B(-p,s)\mathrm{d}p = -[\hat{e}^i(s)\sin(\theta)/s^2 p_0^2 \Delta]$$

$$\{\exp[-sp_0(\ell/2 - z_S + \Delta)] - 2\exp[-sp_0(\ell/2 - z_S)]$$

$$+ \exp[-sp_0(\ell/2 - z_S - \Delta)]\} \tag{2.19}$$

for $\{0 \le \theta < \pi/2\}$ and

$$\frac{s}{2i\pi} \int_{p=-i\infty}^{i\infty} \tilde{E}_z^s(0,p,s)\tilde{I}^B(-p,s)\mathrm{d}p = -[\hat{e}^i(s)/2\Delta]$$

$$[(\ell/2 - z_S + \Delta)^2 - 2(\ell/2 - z_S)^2 + (\ell/2 - z_S - \Delta)^2] \tag{2.20}$$

for $\theta = \pi/2$ as $a \downarrow 0$. Both Eqs. (2.19) and (2.20) can be readily transformed to the TD, thus getting the time-dependent elements of the excitation array for the plane-wave excitation (see Eq. (2.15)), that is

$$V^{[S]}(t) = -[\sin(\theta)/p_0^2\Delta]e^i(t)$$

$$*_t \{[t - p_0(\ell/2 - z_S + \Delta)]\mathrm{H}[t - p_0(\ell/2 - z_S + \Delta)]$$

$$- 2[t - p_0(\ell/2 - z_S)]\mathrm{H}[t - p_0(\ell/2 - z_S)]$$

$$+ [t - p_0(\ell/2 - z_S - \Delta)]\mathrm{H}[t - p_0(\ell/2 - z_S - \Delta)]\} \tag{2.21}$$

for all $S = \{1, \ldots, N\}$ with $\{0 \le \theta < \pi/2\}$ and

$$V^{[S]}(t) = -[e^i(t)/2\Delta]$$

$$[(\ell/2 - z_S + \Delta)^2 - 2(\ell/2 - z_S)^2 + (\ell/2 - z_S - \Delta)^2] \tag{2.22}$$

for all $S = \{1, \ldots, N\}$ with $\theta = \pi/2$. In numerical code implementations, the time convolution in Eq. (2.21) can be approximated with the aid of the trapezoidal rule, for example.

2.4.2 Delta-Gap Excitation

The incident-field distribution in a gap, where the antenna is excited through a voltage pulse, can be described by

$$\hat{E}_z^i(a,z,s) = [\hat{V}^T(s)/\delta][\mathrm{H}(z - z_\delta + \delta/2) - \mathrm{H}(z - z_\delta - \delta/2)] \tag{2.23}$$

where z_δ localizes the center of the gap, $\delta > 0$ is its width, and $\hat{V}^T(s)$ denotes the complex-FD counterpart of the excitation voltage pulse. Again, the residue theorem

[30, section 3.11] can be used to show that the transform-domain counterpart of Eq. (2.23) has the form

$$\tilde{E}_z^i(a, p, s) = \hat{V}^T(s) i_0(sp\delta/2) \exp(spz_\delta) \tag{2.24}$$

where $i_0(x)$ denotes the modified spherical Bessel function of the first kind. Making use of the explicit-type boundary condition (2.16) along with Eq. (2.24) in the right-hand side of Eq. (2.6), we get a slowness-domain integral that can be easily evaluated via Cauchy's formula [30, section 2.41]. Upon transforming the result to the TD, we arrive at the following expression specifying the elements of the excitation array for the delta-gap voltage excitation, that is

$$\begin{aligned}
V^{[S]}(t) = &- [V^T(t)/2\delta\Delta][(z_\delta + \Delta + \delta/2 - z_S)^2 \, H(z_\delta + \Delta + \delta/2 - z_S) \\
&- (z_\delta + \Delta - \delta/2 - z_S)^2 \, H(z_\delta + \Delta - \delta/2 - z_S) \\
&- 2(z_\delta + \delta/2 - z_S)^2 \, H(z_\delta + \delta/2 - z_S) \\
&+ 2(z_\delta - \delta/2 - z_S)^2 \, H(z_\delta - \delta/2 - z_S) \\
&+ (z_\delta - \Delta + \delta/2 - z_S)^2 \, H(z_\delta - \Delta + \delta/2 - z_S) \\
&- (z_\delta - \Delta - \delta/2 - z_S)^2 \, H(z_\delta - \Delta - \delta/2 - z_S)]
\end{aligned} \tag{2.25}$$

for all $S = \{1, \ldots, N\}$. If the width of the excitation gap can be neglected with respect to the spatial support of the excitation pulse, then Eq. (2.24) boils down to $\tilde{E}_z^i(a, p, s) = \hat{V}_0(s) \exp(spz_\delta)$, which in turn leads to

$$\begin{aligned}
V^{[S]}(t) = &-[V^T(t)/\Delta][(z_\delta + \Delta - z_S)H(z_\delta + \Delta - z_S) \\
&- 2(z_\delta - z_S)H(z_\delta - z_S) + (z_\delta - \Delta - z_S)H(z_\delta - \Delta - z_S)]
\end{aligned} \tag{2.26}$$

for all $S = \{1, \ldots, N\}$.

ILLUSTRATIVE EXAMPLE

- Make use of the closed-form expression (C.12) to find (an approximation of) the TD input impedance of a short dipole antenna with the triangular electric-current spatial distribution (also known as the Abraham dipole).

Figure 2.3 shows a short current-carrying wire with the postulated triangular electric-current spatial distribution. The inner node is located at $z = 0$, and the

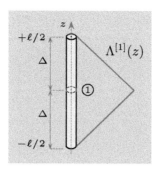

FIGURE 2.3. A short dipole with the triangular electric-current spatial distribution.

length of the discretization segment is equal to half of the wire's length $\ell > 0$. Accordingly, via Eqs. (C.12) with (C.13) for $S = n = 1$, we obtain

$$
\begin{aligned}
Z^{[1,1]}(t) = &- (\zeta_0\ell/\pi c_0 \Delta t) \\
&\times \{(1/6)\{[7/3 + \ln(2)]H(c_0 t - \ell/2) - (14/3)H(c_0 t - \ell)\} \\
&+ (1/6)\ln(c_0 t/\ell)[H(c_0 t - \ell/2) - 2H(c_0 t - \ell)] \\
&- (c_0 t/\ell)^2 \ln(c_0 t/\ell)[2H(c_0 t - \ell/2) - H(c_0 t - \ell)] \\
&- 2(c_0 t/\ell)[H(c_0 t - \ell/2) - H(c_0 t - \ell)] \\
&+ (c_0 t/\ell)^2\{[2 - 2\ln(2)]H(c_0 t - \ell/2) - H(c_0 t - \ell)\} \\
&+ (2/9)(c_0 t/\ell)^3[4H(c_0 t - \ell/2) - H(c_0 t - \ell)] - (2/3)(c_0 t/\ell)^3 H(c_0 t) \\
&+ [(c_0 t/\ell)^2 + 1/6]\cosh^{-1}(c_0 t/a)H(c_0 t - a) - 2(c_0 t/\ell)^2 H(c_0 t)\}
\end{aligned}
\tag{2.27}
$$

In line with the convolution-type Eq. (2.14), in which the electric-current array is related to the excitation voltage array through a central second-order difference (e.g. [25, eq. (25.1.2)]) of the impedance matrix, we express the TD impedance as

$$
Z(t) = Z^{[1,1]}(t + \Delta t) - 2Z^{[1,1]}(t) + Z^{[1,1]}(t - \Delta t)
\tag{2.28}
$$

where Δt denotes the time step along the discretized time axis $\{t_k = k\Delta t; \Delta t > 0, k = 1, 2, \ldots, M\}$. Consequently, the corresponding discrete convolution-type equation has the following form

$$
\sum_{k=1}^{m} Z_{m-k} i_k = V_m
\tag{2.29}
$$

for all $m = \{1, \ldots, M\}$, where we used $Z(t_k) = Z_k$, $i_k = i_k^{[1]}$ and $V_k = V_k^{[1]}$ for brevity. Equation (2.29) can be solved iteratively, which yields the electric-current coefficients excited at $z = 0$ (cf. Eq. (2.15))

$$i_m = Z_0^{-1} \left[V_m - \sum_{k=1}^{m-1} Z_{m-k} i_k \right] \qquad (2.30)$$

for all $m = \{1, \ldots, M\}$.

CHAPTER 3

PULSED EM MUTUAL COUPLING BETWEEN PARALLEL WIRE ANTENNAS

The methodology introduced in chapter 2 can be extended to analyze the pulsed EM interaction between parallel straight wire segments in an antenna array. Such an extension is exactly the main goal of the present chapter. Here, the lines of reasoning closely follow chapter 2. At first, the scattered EM wave fields pertaining to the antenna array are interrelated with the testing EM field state via the EM reciprocity theorem of the time-convolution type. Subsequently, the resulting interaction quantity is approximated through a piecewise linear space-time basis, which results in the electric-current distribution along antenna elements via the marching on-in-time scheme.

3.1 PROBLEM DESCRIPTION

The problem configuration consists of a set of mutually parallel wire segments that can be excited by an incident EM plane wave or/and by delta-gap sources (see section 2.4). An example of the problem under consideration is shown in Figure 3.1. Here, the position is localized via the coordinates $\{x, y, z\}$ with respect to a Cartesian reference frame with the origin $\mathcal{O}$ and the standard basis $\{i_x, i_y, i_z\}$. Again, the antenna system is embedded in the unbounded, homogeneous, loss-free, and isotropic embedding $\mathcal{D}^\infty$ whose EM properties are described by (real-valued, positive and scalar) electric permittivity ϵ_0 and magnetic permeability μ_0. The antenna elements are from the exterior domain separated by sufficiently regular, non-overlapping surfaces, whose union is denoted by $\mathcal{S}_0$. Following the strategy pursued in section 2.1, the scattered EM field accounts for the presence of the antenna array and is defined as the difference between the actual total EM field and the incident EM wave field.

Time-Domain Electromagnetic Reciprocity in Antenna Modeling, First Edition. Martin Štumpf.
© 2020 by The Institute of Electrical and Electronics Engineers, Inc. Published 2020 by John Wiley & Sons, Inc.

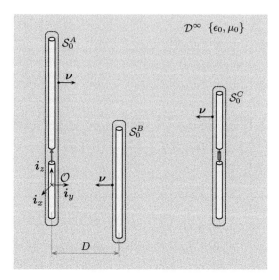

FIGURE 3.1. An antenna array of three parallel wire segments for which $\mathcal{S}_0 = \mathcal{S}_0^A \cup \mathcal{S}_0^B \cup \mathcal{S}_0^C$.

3.2 PROBLEM FORMULATION

Application of the reciprocity theorem to the scattered (s) and testing (B) EM field states and to the unbounded domain exterior to the antenna elements leads to Eq. (2.1), in which $\mathcal{S}_0$ denotes the union of the surfaces enclosing the antenna elements (see Figure 3.1). Under the thin-wire assumption, the reciprocity relation is next approximated by the corresponding on-axis interaction quantity consisting of (the sum of) self and mutual interaction terms. Considering the EM interaction between two PEC antennas denoted by (A) and (B), for example, the interaction quantity has the following form

$$\int_{z=-\ell^A/2}^{\ell^A/2} [\hat{E}_z^B(a^A, z, s) + \hat{E}_z^B(D, z, s)]\hat{I}^{s;A}(z, s)\mathrm{d}z$$

$$+ \int_{z=-\ell^B/2}^{\ell^B/2} [\hat{E}_z^B(a^B, z, s) + \hat{E}_z^B(D, z, s)]\hat{I}^{s;B}(z, s)\mathrm{d}z$$

$$= -\int_{z=-\ell^A/2}^{\ell^A/2} \hat{E}_z^i(a^A, z, s)\hat{I}^{B;A}(z, s)\mathrm{d}z$$

$$- \int_{z=-\ell^B/2}^{\ell^B/2} \hat{E}_z^i(a^B, z, s)\hat{I}^{B;B}(z, s)\mathrm{d}z \tag{3.1}$$

where $\hat{I}^{\mathrm{s};A}$ and $\hat{I}^{\mathrm{s};B}$ denote the (unknown) induced electric current along antennas (A) and (B), respectively, $a^{A,B}$ denotes their radii, and D is the horizontal offset between the interacting antennas. Recall that the relation between testing electric field $\hat{E}_z^{\mathrm{B}}$ is related to the testing current according to Eq. (2.4). An inspection of Eqs. (2.3) and (3.1) reveals that the evaluation of the mutual EM coupling between parallel wire antennas necessitates the handling of a new interaction term, namely

$$\int_{z=-\ell/2}^{\ell/2} \hat{E}_z^{\mathrm{B}}(D, z, s)\hat{I}^{\mathrm{s}}(z, s)\mathrm{d}z \tag{3.2}$$

for an arbitrary distance $D > 0$. As in section 2.2, we shall next tackle the problem via the slowness domain, in which Eq. (3.2) has the following form

$$\frac{s}{2\mathrm{i}\pi}\int_{p=-\mathrm{i}\infty}^{\mathrm{i}\infty} \tilde{E}_z^{\mathrm{B}}(D, p, s)\tilde{I}^{\mathrm{s}}(-p, s)\mathrm{d}p \tag{3.3}$$

and $\tilde{E}_z^{\mathrm{B}}$ is, again, related to $\tilde{I}^{\mathrm{B}}$ via Eq. (2.7).

3.3   PROBLEM SOLUTION

Expanding the desired electric-current distribution along wire elements in the piecewise linear basis functions and assuming the piecewise linear spatial distribution of the testing current (see section 2.3), the interaction quantity describing a two-element antenna array can be cast into a matrix form (see Figure 3.2) whose elements can be directly associated with the interactions given in Eq. (3.1). In

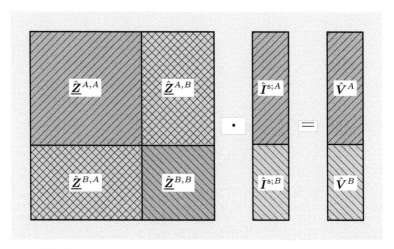

FIGURE 3.2. Impedance matrix description of a two-element antenna array.

the partitioned matrix description, the partial impedance matrix $\hat{\underline{Z}}^{A,A}$ (or $\hat{\underline{Z}}^{B,B}$) represents the interactions between discretization segments on one and the same antenna (A) (or (B)). Its elements can be hence found using the closed-form expressions given in appendix C. Elements of the remaining partial impedance matrices, $\hat{\underline{Z}}^{A,B}$ and $\hat{\underline{Z}}^{B,A}$, represent the remote interaction between the segments on coupled antennas. This remote interaction between two wire antennas is associated with the new interaction term (3.3), whose evaluation leads to the TD mutual-impedance matrix elements. A detailed description of the latter can be found in appendix D.

The excitation of the antenna array is incorporated via the excitation array V that for a two-element antenna array consists of two parts (see Figure 3.2) that correspond to the two terms on the right-hand side of Eq. (3.1). The excitation-array elements can be, in principle, handled along the lines described in section 2.4 for both EM plane-wave and delta-gap excitations. Once the excitation and impedance arrays are filled, we will end up with the system of equations described by Eq. (2.14) whose iterative solution (see Eq. (2.15)) leads to the induced electric-current space-time distribution along the segments of the entire antenna array. Following the described strategy, it is straightforward to extend the partitioned matrix description to a larger system of (mutually parallel) wire antennas.

CHAPTER 4

INCORPORATING WIRE-ANTENNA LOSSES

If the wire antenna is not perfectly conducting, then the effect of losses can be incorporated through the concept of impedance [31]. Using this strategy, the explicit-type boundary condition (2.16) applying to a PEC antenna is generalized by

$$\lim_{r\downarrow a} \hat{E}^{\text{s}}_z(r, z, s) = -\hat{E}^{\text{i}}_z(a, z, s) + \hat{Z}(s)\hat{I}^{\text{s}}(z, s) \tag{4.1}$$

in $\{-\ell/2 \leq z \leq \ell/2\}$, where $\hat{Z}(s)$ is the internal impedance of a solid wire describing the effect of ohmic losses. A closed-form approximation of this parameter has been derived in appendix E. Substituting next the transform-domain counterpart of the boundary condition in the right-hand side of Eq. (2.6), we get a modified slowness-domain reciprocity relation

$$\frac{s}{2\mathrm{i}\pi} \int_{p=-\mathrm{i}\infty}^{\mathrm{i}\infty} \tilde{E}^{\text{s}}_z(a, p, s)\tilde{I}^{\text{B}}(-p, s)\mathrm{d}p$$

$$= -\frac{s}{2\mathrm{i}\pi} \int_{p=-\mathrm{i}\infty}^{\mathrm{i}\infty} \tilde{E}^{\text{i}}_z(a, p, s)\tilde{I}^{\text{B}}(-p, s)\mathrm{d}p$$

$$+ \frac{s\hat{Z}(s)}{2\pi\mathrm{i}} \int_{p=-\mathrm{i}\infty}^{\mathrm{i}\infty} \tilde{I}^{\text{s}}(p, s)\tilde{I}^{\text{B}}(-p, s)\mathrm{d}p \tag{4.2}$$

Clearly, the first term on the right-hand side of Eq. (4.2) has been evaluated in section 2.4 for both plane-wave and delta-gap excitations. The second (new) term that accounts for the effect of losses described by the internal impedance $\hat{Z}(s)$ is hence the subject of the following section.

4.1 MODIFICATION OF THE IMPEDANCE MATRIX

Substitution of the expansions (2.12)–(2.13) in the second term on the right-hand side of Eq. (4.2) reveals that the effect of losses can be accounted for via an additional impedance matrix whose elements follow from (cf. Eqs. (C.1)–(C.2))

$$\hat{R}^{[S,n]}(s) = \frac{c_0 \hat{Z}(s)/s^2}{c_0 \Delta t \Delta^2} \frac{s}{2\mathrm{i}\pi} \int_{p=-\mathrm{i}\infty}^{\mathrm{i}\infty} \frac{\tilde{F}^{[S,n]}(p,s)}{s^4 p^4} \mathrm{d}p \tag{4.3}$$

for all $S = \{1, \ldots, N\}$ and $n = \{1, \ldots, N\}$, where

$$\tilde{F}^{[S,n]}(p,s) = [\exp(2sp\Delta) - 4\exp(sp\Delta)$$
$$+ 6 - 4\exp(-sp\Delta) + \exp(-2sp\Delta)] \exp[sp(z_n - z_S)] \tag{4.4}$$

In line with appendix C, the integration path in the complex p-plane is first indented around the origin at $p = 0$ as shown in Figure C.1. Secondly, Cauchy's formula [30, section 2.41] is applied to evaluate the integrals. Transforming, finally, the result of the integration to the TD, we end up with (cf. Eq. (C.12))

$$R^{[S,n]}(t) = \left[\mathcal{Z}(t)/6\, c_0 \Delta t \Delta^2\right] \left[(z_n - z_S + 2\Delta)^3 \, \mathrm{H}(z_n - z_S + 2\Delta)\right.$$
$$- 4(z_n - z_S + \Delta)^3 \, \mathrm{H}(z_n - z_S + \Delta) + 6(z_n - z_S)^3 \, \mathrm{H}(z_n - z_S)$$
$$- 4(z_n - z_S - \Delta)^3 \, \mathrm{H}(z_n - z_S - \Delta)$$
$$\left. + (z_n - z_S - 2\Delta)^3 \, \mathrm{H}(z_n - z_S - 2\Delta)\right] \tag{4.5}$$

in which $\mathcal{Z}(t)$ denotes TD counterpart of $c_0 \hat{Z}(s)/s^2$. For the internal impedance pertaining to a solid wire with the ohmic loss as given in Eq. (E.10), the TD impedance reads

$$\mathcal{Z}(t) = \frac{1}{\pi a} \left(\frac{\zeta_0}{\sigma}\right)^{1/2} \left(\frac{c_0 t}{\pi}\right)^{1/2} \mathrm{H}(t) \tag{4.6}$$

Finally, the losses are incorporated by replacing the impedance matrix $\underline{Z}$ in the iterative scheme (2.15) with $\underline{Z} - \underline{R}$ along the discretized time axis. Specifically, the electric-current distribution along a lossy wire antenna is found from

$$I_m = \bar{\underline{Z}}_1^{-1} \cdot \left[V_m - \sum_{k=1}^{m-1} \left(\bar{\underline{Z}}_{m-k+1} - 2\bar{\underline{Z}}_{m-k} + \bar{\underline{Z}}_{m-k-1}\right) \cdot I_k\right] \tag{4.7}$$

for all $m = \{1, \ldots, M\}$, where

$$\bar{\underline{Z}} = \underline{Z} - \underline{R} \tag{4.8}$$

is the modified impedance matrix including the ohmic losses in the antenna cylindrical body.

CHAPTER 5

CONNECTING A LUMPED ELEMENT TO THE WIRE ANTENNA

EM scattering and radiation characteristics of an antenna system can be effectively influenced by connecting circuit elements to its accessible ports. To incorporate a lumped circuit element in the CdH-MoM, we shall, as in chapter 4, modify the boundary condition applying to the axial electric field along the wire (cf. Eqs. (2.16) and (4.1)). Following this strategy, we write

$$\lim_{r \downarrow a} \hat{E}_z^s(r, z, s) = -\hat{E}_z^i(a, z, s) + \hat{\zeta}(s)\delta(z - z_\zeta)\hat{I}^s(z, s) \tag{5.1}$$

in $\{-\ell/2 \le z \le \ell/2\}$, where $\hat{\zeta}(s)$ denotes the impedance of the lumped element and $\{-\ell/2 < z_\zeta < \ell/2\}$ defines the point around which the impedance is concentrated. The transform-domain counterpart of the boundary condition (5.1) is next substituted in the right-hand side of the starting reciprocity relation (2.6) and we get

$$\frac{s}{2i\pi} \int_{p=-i\infty}^{i\infty} \tilde{E}_z^s(a, p, s)\tilde{I}^B(-p, s)\mathrm{d}p$$

$$= -\frac{s}{2i\pi} \int_{p=-i\infty}^{i\infty} \tilde{E}_z^i(a, p, s)\tilde{I}^B(-p, s)\mathrm{d}p$$

$$+ \frac{s\hat{\zeta}(s)}{2\pi i} \hat{I}^s(z_\zeta, s) \int_{p=-i\infty}^{i\infty} \tilde{I}^B(-p, s) \exp(spz_\zeta)\mathrm{d}p \tag{5.2}$$

Since the left-hand side and the first term on the right-hand side of Eq. (5.2) have been closely analyzed in sections 2.3 and 2.4, we shall next focus on the remaining interaction term describing the effect of the lumped element.

Time-Domain Electromagnetic Reciprocity in Antenna Modeling, First Edition. Martin Štumpf.
© 2020 by The Institute of Electrical and Electronics Engineers, Inc. Published 2020 by John Wiley & Sons, Inc.

5.1 MODIFICATION OF THE IMPEDANCE MATRIX

Supposing that the lumped element is connected in between two discretization nodes, say $z_\zeta \in [z_Q, z_{Q+1}]$ with $Q = \{0, \ldots, N\}$, we may interpolate the load current between the corresponding nodal currents and write

$$\hat{I}^s(z_\zeta, s) \simeq \sum_{k=1}^{M} \left[\Lambda^{[Q]}(z_\zeta) i_k^{[Q]} + \Lambda^{[Q+1]}(z_\zeta) i_k^{[Q+1]} \right] \hat{\Lambda}_k(s) \qquad (5.3)$$

where the interpolation weights directly follow from Eq. (2.10) and, in virtue of the end conditions (2.9), we take $i_k^{[0]} = i_k^{[N+1]} = 0$. Furthermore, $\hat{\Lambda}_k(s)$ denotes the complex FD counterpart of the triangular function defined in Eq. (2.11). Next, making use of Eqs. (5.3) and (2.13) in the interaction term, we will end up with the new impedance matrix, denoted by $\underline{L}$, whose elements follow from

$$\hat{L}^{[S,n]}(s) = \frac{c_0 \hat{\zeta}(s)/s^2}{c_0 \Delta t \Delta} \left[\Lambda^{[Q]}(z_\zeta) \delta_{n,Q} + \Lambda^{[Q+1]}(z_\zeta) \delta_{n,Q+1} \right]$$

$$\times \frac{s}{2 i \pi} \int_{p=-i\infty}^{i\infty} \frac{\tilde{H}^{[S,n]}(p, s)}{s^2 p^2} \mathrm{d}p \qquad (5.4)$$

for all $S = \{1, \ldots, N\}$ and $n = \{1, \ldots, N\}$, where $\delta_{m,n} = 1$ for $m = n$ and $\delta_{m,n} = 0$ for $m \neq n$ is the Kronecker delta and

$$\tilde{H}^{[S,n]}(p, s) = [\exp(sp\Delta) - 2 + \exp(-sp\Delta)] \exp[sp(z_\zeta - z_S)] \qquad (5.5)$$

Again, according to appendix C, the integration contour in the complex p-plane is first deformed around the origin at $p = 0$, and the integrals are evaluated with the aid of Cauchy's formula [30, section 2.41]. The subsequent transformation to the TD then leads to

$$L^{[S,n]}(t) = [F(t)/\Delta] \left[\Lambda^{[Q]}(z_\zeta) \delta_{n,Q} + \Lambda^{[Q+1]}(z_\zeta) \delta_{n,Q+1} \right]$$

$$\times \left[(z_\zeta - z_S + \Delta) \, \mathrm{H}(z_\zeta - z_S + \Delta) - 2(z_\zeta - z_S) \, \mathrm{H}(z_\zeta - z_S) \right.$$

$$\left. + (z_\zeta - z_S - \Delta) \, \mathrm{H}(z_\zeta - z_S - \Delta) \right] \qquad (5.6)$$

in which $F(t)$ denotes the TD counterpart of $\hat{\zeta}(s)/s^2 \, \Delta t$ that reads

$$F(t) = R \, (c_0 t/c_0 \Delta t) \, \mathrm{H}(t) \qquad (5.7)$$

for a resistor of resistance R and

$$F(t) = (\Delta t/2C)(c_0 t/c_0 \Delta t)^2 \, \mathrm{H}(t) \qquad (5.8)$$

for a capacitor of capacitance C, and finally

$$F(t) = (L/\Delta t)\, \mathrm{H}(t) \tag{5.9}$$

for an inductor of inductance L. Like in section 4.1, the lumped element is finally included in by replacing the impedance matrix $\underline{Z}$ in the iterative procedure (2.15) with $\underline{Z} - \underline{L}$ along the discretized time axis. Specifically, the induced space-time electric-current distribution is found from Eq. (4.7) for which we define the modified impedance matrix, that is

$$\overline{\underline{Z}} = \underline{Z} - \underline{L} \tag{5.10}$$

Finally note that it is straightforward to apply the present description to the inclusion of a number of lumped elements and that a serial RLC network connected at $z = z_\zeta$ corresponds to a mere superposition of expressions (5.7)–(5.9).

CHAPTER 6

PULSED EM RADIATION FROM A STRAIGHT WIRE ANTENNA

The main result of chapter 2 is the induced electric-current space-time distribution along a wire antenna which is sufficient to obtain all its EM scattering and radiation characteristics. To facilitate the calculation of such characteristics, we next provide formulas expressing the radiated EM fields from a straight wire segment in terms of the on-axis electric-current distribution. The ensuing analysis relies heavily on the source-type representations for EM fields radiated from EM sources of bounded extent in an unbounded homogeneous, isotropic medium [3, chapter 26] that are adapted to a straight, thin-wire radiator. Their straightforward numerical handling via the trapezoidal rule is proposed.

6.1 PROBLEM DESCRIPTION

We shall analyze EM radiation from a straight wire segment shown in Figure 6.1. Position in the problem configuration is localized through the position vector $r = x i_x + y i_y + z i_z$, where $\{x, y, z\}$ are coordinates defined with respect to a Cartesian reference frame with the origin $\mathcal{O}$ and the standard basis $\{i_x, i_y, i_z\}$. Again, the antenna is placed in an unbounded, homogeneous, loss-free, and isotropic embedding $\mathcal{D}^\infty$ whose EM properties are described by electric permittivity ϵ_0 and magnetic permeability μ_0. The corresponding EM wave speed is $c_0 = (\epsilon_0 \mu_0)^{-1/2}$ and $\zeta_0 = (\mu_0/\epsilon_0)^{1/2}$ denotes the wave impedance. Under these assumptions, the radiated EM field is governed by the free-space EM field equations [3, section 18.2] whose explicit solution for a thin, straight wire radiator is given in the following section.

Time-Domain Electromagnetic Reciprocity in Antenna Modeling, First Edition. Martin Štumpf.

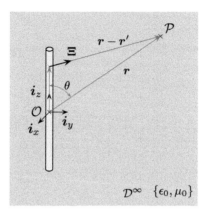

FIGURE 6.1. A straight wire segment configuration.

6.2   SOURCE-TYPE REPRESENTATIONS FOR THE TD RADIATED EM FIELDS

Assuming the radiating electric-current source distributed along the wire axis with support in $\{-\ell/2 < z < \ell/2\}$, the general source-type representations (see [3, section 26.4]) for the electric-field strength can be expressed as a sum of the near-field (NF), intermediate-field (IF), and far-field (FF) terms, that is

$$\boldsymbol{E}(r,z,t) = \boldsymbol{E}^{\mathrm{NF}}(r,z,t) + \boldsymbol{E}^{\mathrm{IF}}(r,z,t) + \boldsymbol{E}^{\mathrm{FF}}(r,z,t) \tag{6.1}$$

in which

$$\boldsymbol{E}^{\mathrm{NF}}(r,z,t) = -\epsilon_0^{-1} \int_{\zeta=-\ell/2}^{\ell/2} \{\boldsymbol{i}_z - 3[\boldsymbol{i}_z \cdot \boldsymbol{\Xi}(r,z|\zeta)]\boldsymbol{\Xi}(r,z|\zeta)\}$$
$$\frac{\partial_t^{-1} I^{\mathrm{s}}\{\zeta, t - [r^2 + (z-\zeta)^2]^{1/2}/c_0\}}{4\pi[r^2 + (z-\zeta)^2]^{3/2}} \mathrm{d}\zeta \tag{6.2}$$

$$\boldsymbol{E}^{\mathrm{IF}}(r,z,t) = -\zeta_0 \int_{\zeta=-\ell/2}^{\ell/2} \{\boldsymbol{i}_z - 3[\boldsymbol{i}_z \cdot \boldsymbol{\Xi}(r,z|\zeta)]\boldsymbol{\Xi}(r,z|\zeta)\}$$
$$\frac{I^{\mathrm{s}}\{\zeta, t - [r^2 + (z-\zeta)^2]^{1/2}/c_0\}}{4\pi[r^2 + (z-\zeta)^2]} \mathrm{d}\zeta \tag{6.3}$$

$$\boldsymbol{E}^{\mathrm{FF}}(r,z,t) = -\mu_0 \int_{\zeta=-\ell/2}^{\ell/2} \{\boldsymbol{i}_z - [\boldsymbol{i}_z \cdot \boldsymbol{\Xi}(r,z|\zeta)]\boldsymbol{\Xi}(r,z|\zeta)\}$$
$$\frac{\partial_t I^{\mathrm{s}}\{\zeta, t - [r^2 + (z-\zeta)^2]^{1/2}/c_0\}}{4\pi[r^2 + (z-\zeta)^2]^{1/2}} \mathrm{d}\zeta \tag{6.4}$$

where the unit vector in the direction of observation is found from (see Figure 6.1)

$$\mathbf{\Xi}(r, z|\zeta) = \frac{\mathbf{r} - \mathbf{r}'}{|\mathbf{r} - \mathbf{r}'|} = \frac{r\mathbf{i}_r + (z - \zeta)\mathbf{i}_z}{[r^2 + (z - \zeta)^2]^{1/2}} \tag{6.5}$$

with $r = (x^2 + y^2)^{1/2} > 0$ and $r\mathbf{i}_r = x\mathbf{i}_x + y\mathbf{i}_y$. Similar expressions apply to the radiated magnetic-field strength. Since the near-field component of the latter is identically zero, we have

$$\mathbf{H}(r, z, t) = \mathbf{H}^{\text{IF}}(r, z, t) + \mathbf{H}^{\text{FF}}(r, z, t) \tag{6.6}$$

where

$$\mathbf{H}^{\text{IF}}(r, z, t) = \int_{\zeta=-\ell/2}^{\ell/2} \{\mathbf{i}_z \times \mathbf{\Xi}(r, z|\zeta)\}$$

$$\frac{I^s\{\zeta, t - [r^2 + (z - \zeta)^2]^{1/2}/c_0\}}{4\pi[r^2 + (z - \zeta)^2]} d\zeta \tag{6.7}$$

$$\mathbf{H}^{\text{FF}}(r, z, t) = c_0^{-1} \int_{\zeta=-\ell/2}^{\ell/2} \{\mathbf{i}_z \times \mathbf{\Xi}(r, z|\zeta)\}$$

$$\frac{\partial_t I^s\{\zeta, t - [r^2 + (z - \zeta)^2]^{1/2}/c_0\}}{4\pi[r^2 + (z - \zeta)^2]^{1/2}} d\zeta \tag{6.8}$$

The directional patterns become perhaps more transparent in the spherical coordinate system $\{R, \theta, \phi\}$ for $\zeta = 0$, that is

$$\mathbf{i}_z - (\mathbf{i}_z \cdot \mathbf{\Xi})\mathbf{\Xi}|_{\zeta=0} = -\mathbf{i}_\theta \sin(\theta) \tag{6.9}$$

$$\mathbf{i}_z - 3(\mathbf{i}_z \cdot \mathbf{\Xi})\mathbf{\Xi}|_{\zeta=0} = -\mathbf{i}_\theta \sin(\theta) - 2\mathbf{i}_R \cos(\theta) \tag{6.10}$$

$$\mathbf{i}_z \times \mathbf{\Xi}|_{\zeta=0} = \mathbf{i}_\phi \sin(\theta) \tag{6.11}$$

which apply to a short-wire antenna oriented along the z-direction [3, section 26.9].

In numerical calculations, we take into account the piecewise linear distribution of the induced electric-current distribution as assumed in the chosen solution methodology (see Figure 2.2b) and apply the trapezoidal rule approximation to the spatial integrations in Eqs. (6.2)–(6.4), (6.7), and (6.8). The z-component of the electric-field strength, for example, can be then written as

$$E_z^{\text{NF}}(r, z, t) = -\epsilon_0^{-1}\Delta \sum_{n=1}^{N} \{1 - 3[\mathbf{i}_z \cdot \mathbf{\Xi}(r, z|\zeta_n)]^2\}$$

$$\frac{\partial_t^{-1} i^{[n]}\{t - [r^2 + (z - \zeta_n)^2]^{1/2}/c_0\}}{4\pi[r^2 + (z - \zeta_n)^2]^{3/2}} \tag{6.12}$$

$$E_z^{\mathrm{IF}}(r, z, t) = -\zeta_0 \Delta \sum_{n=1}^{N} \{1 - 3[\mathbf{i}_z \cdot \mathbf{\Xi}(r, z|\zeta_n)]^2\}$$

$$\frac{i^{[n]}\{t - [r^2 + (z - \zeta_n)^2]^{1/2}/c_0\}}{4\pi[r^2 + (z - \zeta_n)^2]} \qquad (6.13)$$

$$E_z^{\mathrm{FF}}(r, z, t) = -\mu_0 \Delta \sum_{n=1}^{N} \{1 - [\mathbf{i}_z \cdot \mathbf{\Xi}(r, z|\zeta_n)]^2\}$$

$$\frac{\partial_t\, i^{[n]}\{t - [r^2 + (z - \zeta_n)^2]^{1/2}/c_0\}}{4\pi[r^2 + (z - \zeta_n)^2]^{1/2}} \qquad (6.14)$$

where $i^{[n]}$ is the induced electric-current pulse at the n-th discretization node, $\zeta_n = -\ell/2 + n\Delta$ and N denotes the number of inner spatial-discretization points (see Figure 2.2a). Owing to the fact that the calculated electric-current pulses are also piecewise linear functions in time, the temporal integration and differentiation that appear in Eqs. (6.12) and (6.14) can be carried out analytically. Note in this respect that the trapezoidal rule leads to the exact result for any piecewise linear function.

6.3   FAR-FIELD TD RADIATION CHARACTERISTICS

The complex FD electric-field strength radiated from a z-oriented, straight thin-wire segment in an unbounded, homogeneous, and isotropic embedding can be found from [3, eq. (26.9-2)]

$$\hat{\mathbf{E}}(r, z, s) = -s\mu_0 \hat{\Phi}_z(r, z, s)\mathbf{i}_z + (s\epsilon_0)^{-1}\, \nabla\, \partial_z \hat{\Phi}_z(r, z, s) \qquad (6.15)$$

where we used $\nabla = \partial_x \mathbf{i}_x + \partial_y \mathbf{i}_y + \partial_z \mathbf{i}_z$ (see section 1.2), and

$$\hat{\Phi}_z(r, z, s) = \int_{\zeta=-\ell/2}^{\ell/2} \frac{\exp\{-s[r^2 + (z - \zeta)^2]^{1/2}/c_0\}}{4\pi[r^2 + (z - \zeta)^2]^{1/2}} \hat{I}^{\mathrm{s}}(\zeta, s)\mathrm{d}\zeta \qquad (6.16)$$

is the z-component of the complex-FD electric-current vector potential. Using Taylor's expansion about $R = (r^2 + z^2)^{1/2} \to \infty$, the latter can be written as

$$\hat{\Phi}_z(r, z, s) = \hat{\Phi}_z^{\infty}(\theta, s)\frac{\exp(-sR/c_0)}{4\pi R}[1 + O(1/R)] \text{ as } R \to \infty \qquad (6.17)$$

where

$$\hat{\Phi}_z^{\infty}(\theta, s) = \int_{\zeta=-\ell/2}^{\ell/2} \hat{I}^{\mathrm{s}}(\zeta, s) \exp[s\zeta \cos(\theta)/c_0]\mathrm{d}\zeta \qquad (6.18)$$

and $\cos(\theta) = z/R$. Consequently, upon expanding the electric-field strength, that is

$$\hat{E}(r, z, s) = \hat{E}^\infty(\theta, s) \frac{\exp(-sR/c_0)}{4\pi R} [1 + O(1/R)] \text{ as } R \to \infty \qquad (6.19)$$

we may use Eqs. (6.18) and (6.19) in Eq. (6.15) to express the amplitude electric-field radiation characteristics $\hat{E}^\infty(\theta, s)$. Transforming the result to the TD, the polar component of the radiation characteristic reads

$$E_\theta^\infty(\theta, t) = \mu_0 \partial_t \Phi_z^\infty(\theta, t) \sin(\theta) \qquad (6.20)$$

in which the TD counterpart of Eq. (6.18) follows

$$\Phi_z^\infty(\theta, t) = \int_{\zeta=-\ell/2}^{\ell/2} I^s[\zeta, t + \zeta \cos(\theta)/c_0] \mathrm{d}\zeta \qquad (6.21)$$

Again, as the calculated electric-current distribution is a piecewise linear function (see Figure 2.2a), the integration in Eq. (6.21) can be calculated with the aid of the trapezoidal rule as

$$\Phi_z^\infty(\theta, t) = \Delta \sum_{n=1}^{N} i^{[n]}[t + \zeta_n \cos(\theta)/c_0] \qquad (6.22)$$

where we have accounted for the end conditions (2.9). Recall that $\zeta_n = -\ell/2 + n\Delta$ and $i^{[n]}$ represents the induced electric-current pulse at the n-th discretization node. Since the latter directly follows from the iterative procedure (2.15), it is straightforward to calculate the pulsed EM radiation characteristics at any observation angle θ using Eq. (6.22) in Eq. (6.20).

CHAPTER 7

EM RECIPROCITY BASED CALCULATION OF TD RADIATION CHARACTERISTICS

A systematic use of the EM reciprocity theorem of the time-convolution type results in the (self-)reciprocity relation facilitating the construction of equivalent Kirchhoff's network antenna representations (see [28, 32] and [4, chapter 5], for example). This relation can also be applied to calculating the TD EM radiation characteristics of an antenna system from its load response to a uniform EM plane wave in reception (e.g. [33]). This is exactly the main goal of this chapter, where such a reciprocity relation concerning a wire antenna is discussed and numerically validated.

7.1 PROBLEM DESCRIPTION

In this chapter, the transmitting (T) and receiving (R) states of one and the same wire antenna are mutually interrelated (see Figure 7.1). In the transmitting situation, the antenna is activated by a voltage pulse $V^T(t)$ applied in its excitation gap. Consequently, the antenna radiates into its embedding, where we may calculate the TD far-field radiated amplitude at a given angle of observation θ. The TD radiation characteristics can be calculated either from the electric-current distribution along the radiating wire (see section 6.3) or, alternatively, using the induced load voltage and current quantities in the receiving state in which the antenna is irradiated by an impulsive EM plane wave. The latter is defined by (see Eq. (11.28))

- $e^i(t) =$ plane-wave signature;
- $\alpha =$ a unit vector in the direction of polarization;
- $\beta =$ a unit vector in the direction of propagation.

Time-Domain Electromagnetic Reciprocity in Antenna Modeling, First Edition. Martin Štumpf.

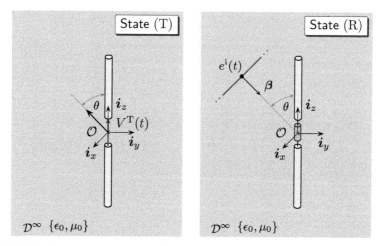

FIGURE 7.1. Transmitting and receiving scenarios of one and the same wire antenna.

Owing to the rotational symmetry of the problem configuration (see Figure 7.1), the polarization and propagation vectors can be specified through the polar angle by $\alpha = i_x \cos(\theta) + i_z \sin(\theta)$ and $\beta = i_x \sin(\theta) - i_z \cos(\theta)$. Likewise, the unit vector in the polar direction at $\phi = 0$ can be expressed as $i_\theta = i_x \cos(\theta) - i_z \sin(\theta)$, which will be used to express the co-polarized component of the radiated electric far-field amplitude. Again, the antenna under consideration is placed in the unbounded, homogeneous, loss-free, and isotropic embedding $\mathcal{D}^\infty$.

7.2   PROBLEM SOLUTION

The point of departure is the TD counterpart of the reciprocity relation [4, eq. (5.6)] applying to a one-port antenna, that is

$$V^R(t) *_t I^T(t) + V^T(t) *_t I^R(t)$$
$$= \mu_0^{-1} \partial_t^{-1} e^i(t) *_t \alpha \cdot E^{T;\infty}(-\beta, t) \tag{7.1}$$

where

- $V^T(t)$ = the excitation voltage pulse in the gap of the antenna in state (T);
- $I^T(t)$ = the electric-current response of the transmitting wire antenna at the position of its voltage-gap excitation;
- $E^{T;\infty}(\xi, t)$ = electric-field (vectorial) amplitude radiation characteristics of the antenna observed at the direction specified by a unit vector of observation ξ;
- $e^i(t)$ = the incident plane-wave signature in state (R);

- $V^{\mathrm{R}}(t) =$ the voltage across the antenna load;
- $I^{\mathrm{R}}(t) =$ the electric current flowing across the antenna load.

Now, if the plane-wave signature is related to the excitation voltage pulse via

$$e^{\mathrm{i}}(t) = c_0^{-1} \partial_t V^{\mathrm{T}}(t) \tag{7.2}$$

we may rewrite Eq. (7.1) to the following form (see section 6.3)

$$V^{\mathrm{R}}(t) *_t I^{\mathrm{T}}(t) + V^{\mathrm{T}}(t) *_t I^{\mathrm{R}}(t) = -\zeta_0^{-1} V^{\mathrm{T}}(t) *_t E_\theta^{\mathrm{T};\infty}(\theta, t) \tag{7.3}$$

where $E_\theta^{\mathrm{T};\infty}(\theta, t)$ denotes the polar component of the radiated electric-field far-field amplitude in the direction of observation specified by θ, and $\zeta_0 = (\mu_0/\epsilon_0)^{1/2}$ denotes the wave impedance. The reciprocity relation (7.3) will next be assessed numerically with the help of sections 2.4.1 and 5.1.

ILLUSTRATIVE NUMERICAL EXAMPLE

- Make use of the reciprocity relation (7.3) to calculate the TD far-field amplitude in a chosen direction of observation θ from its TD load response induced by an incident EM plane wave. Subsequently, validate the result by calculating the radiated EM pulse directly from the electric-current distribution according to section 6.3.

Solution: We shall analyze TD EM radiation from a wire antenna at $\theta = \pi/8$. The antenna length is $\ell = 0.10\,\mathrm{m}$, and its radius is $a = 0.10\,\mathrm{mm}$. The antenna is excited by a voltage pulse applied in a narrow gap placed at its center $z_\delta = 0$. The excitation voltage pulse shape is described by

$$V^{\mathrm{T}}(t) = 2V_{\mathrm{m}} \left[\left(\frac{t}{t_{\mathrm{w}}}\right)^2 H(t) - 2\left(\frac{t}{t_{\mathrm{w}}} - \frac{1}{2}\right)^2 H\left(\frac{t}{t_{\mathrm{w}}} - \frac{1}{2}\right) \right.$$
$$\left. +2\left(\frac{t}{t_{\mathrm{w}}} - \frac{3}{2}\right)^2 H\left(\frac{t}{t_{\mathrm{w}}} - \frac{3}{2}\right) - \left(\frac{t}{t_{\mathrm{w}}} - 2\right)^2 H\left(\frac{t}{t_{\mathrm{w}}} - 2\right) \right] \tag{7.4}$$

where we take $V_{\mathrm{m}} = 1.0\,\mathrm{V}$ and $c_0 t_{\mathrm{w}} = 1.0\,\ell$. It is straightforward to verify that the time derivative of $V^{\mathrm{T}}(t)$ has the shape of a bipolar triangular pulse. The excitation voltage pulse and the corresponding plane-wave signature calculated using Eq. (7.2) are shown in Figure 7.2a and b, respectively.

Once the excitation voltage pulse and the EM plane-wave signature are defined, we may calculate the electric-current response $I^{\mathrm{T}}(t)$ in transmission

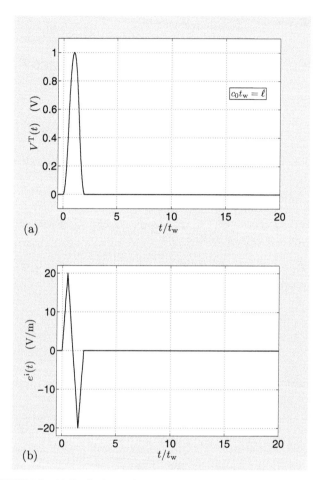

(a)

(b)

FIGURE 7.2. (a) Excitation voltage pulse shape; (b) plane-wave signature.

and $\{V^{\mathrm{R}}, I^{\mathrm{R}}\}(t)$ in reception. Clearly, the relation between the latter quantities depends on the character of the chosen lumped element. In the present example, we take a purely resistive load with $R = 50\ \Omega$. Consequently, the load voltage response is just a scaled copy of the calculated current response at $z_\zeta = 0$, that is, $V^{\mathrm{R}}(t) = R\ I^{\mathrm{R}}(t)$, and its plot can be hence omitted. The electric-current pulses calculated according to chapters 2 and 5 are shown in Figure 7.3. Apparently, the induced current flowing across the load does not start at $t = 0$, which is caused by our choice of the reference of the incident plane wave that hits the top end of the antenna at $t = 0$ (see section 2.4.1). For the sake of validation, this time shift that amounts to $(\ell/2c_0) \cos(\theta)$ will be further compensated.

With the load response at our disposal, we may next evaluate the time convolutions on the left-hand side of Eq. (7.3), which is proportional to

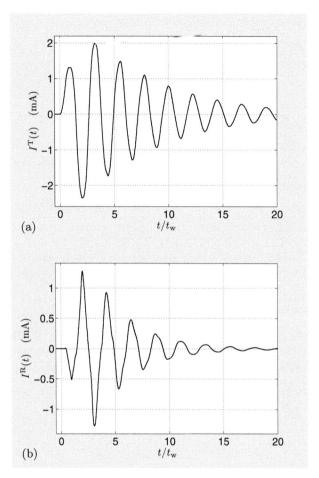

FIGURE 7.3. Electric-current response at $z = 0$. (a) Response to the excitation gap voltage; (b) response to the incident plane wave.

$Q(t) = V^{\mathrm{T}}(t) *_t E_\theta^{\mathrm{T};\infty}(\theta, t)$. To find the far-field amplitude, it is still necessary to perform deconvolution. Fortunately, for the excitation voltage pulse given in Eq. (7.4), this step can be carried out analytically using the sum of (a finite number of) shifted copies of $\partial_t^3 Q(t)$, that is

$$E_\theta^{\mathrm{T};\infty}(\theta, t) = \frac{t_{\mathrm{w}}^2}{4V_{\mathrm{m}}} \sum_{n=0}^{\infty} \frac{2(n+2)^2 - 1 + (-1)^{n+2}}{8} \partial_t^3 Q(t - nt_{\mathrm{w}}/2) \qquad (7.5)$$

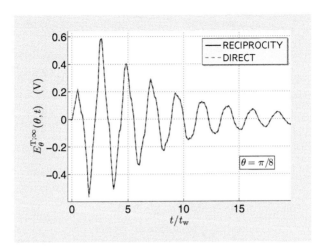

FIGURE 7.4. The TD radiated far-field amplitude at $\theta = \pi/8$ as calculated via EM reciprocity ("RECIPROCITY") and directly ("DIRECT") from the induced electric-current distribution.

The thus obtained far-field amplitude is plotted in Figure 7.4 ("RECIPROCITY") along with the result calculated from the electric-current distribution according to section 6.3 ("DIRECT"). The calculated radiated pulses are almost on top of each, thereby validating the proposed modeling methodologies.

CHAPTER 8

INFLUENCE OF A WIRE SCATTERER ON A TRANSMITTING WIRE ANTENNA

The antenna performance may be essentially influenced via EM coupling between the antenna and its surrounding objects. It is therefore of high importance to quantify these EM coupling effects in terms of observable parameters characterizing the antenna itself. This is exactly the main purpose of [4, chapter 6], where the impact of a scatterer on a multi-port antenna system is described in terms of the corresponding equivalent Kirchhoff-network quantities. A result from this category is also the main subject of the present chapter, where the effect of a straight wire segment on the response of a transmitting wire antenna is analyzed and subsequently numerically evaluated with the aid of methodologies introduced in chapters 3 and 6.

8.1 PROBLEM DESCRIPTION

We shall analyze the impact of a thin-wire PEC scatterer on the impulsive electric-current response $I^T(t)$ of a thin-wire antenna oriented parallel to the scatterer. For this reason, we make use of the reciprocity theorem of the time-convolution type, again, to interrelate two EM field states, say (T) and ($\tilde{\text{T}}$), differing from each other in the presence of a wire scatterer (see Figure 8.1). The horizontal offset between the antenna and the scatterer is denoted by $D > 0$. The wire antenna is in both scenarios excited via one and the same voltage pulse, $V^T(t)$, applied in a narrow gap in the antenna body. For the sake of simplicity, both the antenna and the scatterer are placed in the unbounded, homogeneous, loss-free, and isotropic embedding $\mathcal{D}^\infty$.

Time-Domain Electromagnetic Reciprocity in Antenna Modeling, First Edition. Martin Štumpf.

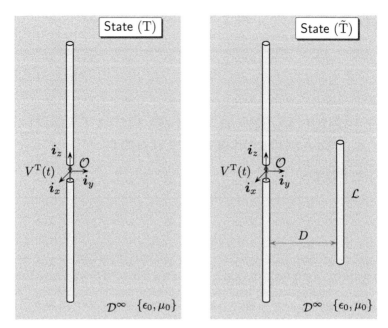

FIGURE 8.1. Transmitting scenarios differing from each other in the presence of a wire scatterer.

8.2 PROBLEM SOLUTION

We start from the TD counterpart of the reciprocity relation [4, eq. (6.27)] that is first adapted to a thin-wire scatterer extending along a line $\mathcal{L}$, that is

$$
\Delta V^{\mathrm{T}}(t) *_t I^{\mathrm{T}}(t) - V^{\mathrm{T}}(t) *_t \Delta I^{\mathrm{T}}(t)
$$
$$
= - \int_{\mathcal{L}} E_z^{\mathrm{T}}(D, \zeta, t) *_t I^{\tilde{s}}(D, \zeta, t) \mathrm{d}\zeta \tag{8.1}
$$

where

- $\Delta I^{\mathrm{T}}(t) = I^{\tilde{\mathrm{T}}}(t) - I^{\mathrm{T}}(t)$ = the change of the electric-current antenna response that we seek;
- $\Delta V^{\mathrm{T}}(t) = V^{\tilde{\mathrm{T}}}(t) - V^{\mathrm{T}}(t)$ = the change of the excitation pulse. As the wire antenna is in the both transmitting situations excited by one and the same voltage pulse, this change is identically zero.
- $E_z^{\mathrm{T}}(D, \zeta, t)$ = the z-component of the total electric field along $\mathcal{L}$ in the *absence* of the wire scatterer. The radiated field can be calculated through formulas (6.12)–(6.14);
- $I^{\tilde{s}}(D, \zeta, t)$ = the axial electric current induced along the scatterer in state $(\tilde{\mathrm{T}})$.

Consequently, making use of $\Delta V^{\mathrm{T}}(t) = 0$ in Eq. (8.1), we will end up with the desired reciprocity relation

$$V^{\mathrm{T}}(t) *_t \Delta I^{\mathrm{T}}(t) = \int_{\mathcal{L}} E_z^{\mathrm{T}}(D,\zeta,t) *_t I^{\tilde{s}}(D,\zeta,t)\mathrm{d}\zeta \tag{8.2}$$

Since the required electric-current distribution $I^{\tilde{s}}(\zeta,t)$ along the wire scatterer $\mathcal{L}$ will be calculated, in line with chapter 3, at (a finite number of) its nodal discretization points only, we shall next apply the trapezoidal rule of integration to approximate the reciprocity relation by

$$V^{\mathrm{T}}(t) *_t \Delta I^{\mathrm{T}}(t) \simeq \Delta \sum_{n=1}^{N} E_z^{\mathrm{T}}(D,\zeta_n,t) *_t I^{\tilde{s}}(D,\zeta_n,t) \tag{8.3}$$

where N denotes the number of the discretization nodes along the scatterer, and Δ is the corresponding grid spacing. The reciprocity relation will be next assessed numerically by comparing $\Delta I^{\mathrm{T}}(t)$ as calculated from the right-hand side of Eq. (8.3) with the corresponding result found directly from the current difference $I^{\tilde{\mathrm{T}}}(t) - I^{\mathrm{T}}(t)$.

ILLUSTRATIVE NUMERICAL EXAMPLE

- Make use of the reciprocity relation (8.2) to calculate the impact of a PEC wire scatterer on the TD electric-current response of a gap-excited wire antenna. Subsequently, validate the result by calculating the change of the response directly from the electric-current distributions calculated according to sections 2.3 and 3.3.

Solution: We shall analyze the impact of a PEC thin-wire scatterer placed along $\{r = D = 0.050\,\mathrm{m}, -0.075\,\mathrm{m} < z < 0.075\,\mathrm{m}\}$ on the transmitting antenna extending along the z-axis at $\{r = 0, -0.050\,\mathrm{m} < z < 0.050\,\mathrm{m}\}$. Both wire structures have a radius $a = 0.10\,\mathrm{mm}$. The transmitting antenna is at its center $z = z_\delta = 0$ excited by a voltage pulse defined by Eq. (7.4) with $V_{\mathrm{m}} = 1.0\,\mathrm{V}$ and $c_0 t_{\mathrm{w}} = 0.10\,\mathrm{m}$. The corresponding pulse shape is shown in Figure 7.2a.

In the first step, we calculate $E_z^{\mathrm{T}}(D,\zeta_n,t)$ along the spatial grid of the wire scatter. This is accomplished by calculating the induced electric-current distribution along the transmitting antenna (see section 2.3), which is subsequently used as the input for the evaluation of radiated pulses through Eqs. (6.12)–(6.14). Secondly, we follow the approach introduced in chapter 3 and evaluate the space-time electric-current distribution $I^{\tilde{s}}(D,\zeta_n,t)$ along the discrete nodes on the coupled wire scatterer. Examples of the radiated electric-field and induced electric-current

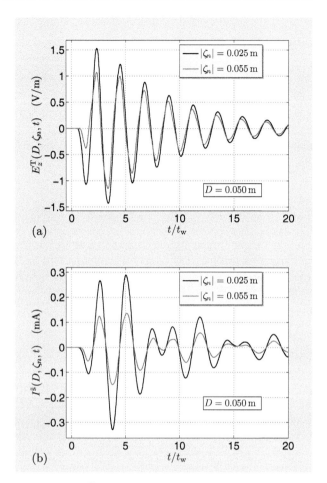

FIGURE 8.2. (a) Radiated E_z^{T} pulses in absence of the scatterer; (b) induced electric-current pulses on the scatterer.

pulse shapes at two points on the scattering wire are shown in Figure 8.2a and b, respectively.

With the set of the radiated electric-field and induced electric-current pulses at our disposal, we can evaluate the right-hand side of Eq. (8.3) and get, say, $P(t) = V^{\mathrm{T}}(t) *_t \Delta I^{\mathrm{T}}(t)$. The latter apparently calls for a deconvolution algorithm, an example of which has been previously applied in section 7.2. Along these lines, we write

$$\Delta I^{\mathrm{T}}(t) = \frac{t_{\mathrm{w}}^2}{4V_{\mathrm{m}}} \sum_{n=0}^{\infty} \frac{2(n+2)^2 - 1 + (-1)^{n+2}}{8} \partial_t^3 P(t - nt_{\mathrm{w}}/2) \qquad (8.4)$$

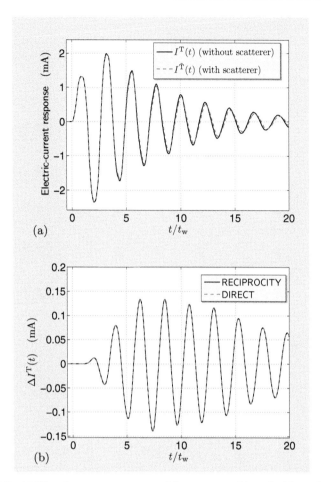

FIGURE 8.3. (a) Electric-current response of the antenna with and without the scatterer; (b) the change of the electric-current response.

which directly yields the desired change of the electric-current response at $z = 0$ (see "RECIPROCITY" in Figure 8.3b). To validate the reciprocity-based methodology, we have further analyzed the both transmitting scenarios shown in Figure 8.1. The electric-current response $I^T(t)$ has been calculated using the methodology described in section 2.3, and the electric current in situation $(\tilde{T})$ has been found with the help of the extension described in section 3.3. The resulting electric-current–pulsed responses along with their difference (see "DIRECT" in Figure 8.3b) are shown in Figure 8.3a and b, respectively. As can be observed, the pulse shapes of $\Delta I^T(t)$ as evaluated with the aid of the reciprocity relation (8.2) and directly by calculating $I^{\tilde{T}}(t) - I^T(t)$ overlap each other. This correspondence proves the consistency of the modeling methodologies presented in chapters 2, 3, and 6.

CHAPTER 9

INFLUENCE OF A LUMPED LOAD ON EM SCATTERING OF A RECEIVING WIRE ANTENNA

EM scattering of a receiving antenna can be affected by changing the impedance of a lumped load connected to its accessible ports. In [4, chapter 9], the EM reciprocity theorem of the time-convolution type is applied to show that the change in such antenna's EM scattering characteristics is intimately related to the corresponding transmitting state, namely, to antenna's (impulse-excited) radiation characteristics. This EM reciprocity relation is also the subject of the present chapter, where we evaluate the change of EM scattering of a wire antenna due to the change in its lumped load. At first, we employ the reciprocity relation and find the impact of a variable load from the relevant radiated far-field amplitude and the corresponding voltage difference across the load. Secondly, for validation purposes, the change of EM scattering characteristics is evaluated directly with the aid of results discussed in sections 2.4.1 and 6.3.

9.1 PROBLEM DESCRIPTION

In this chapter, we shall analyze EM plane-wave scattering of two receiving scenarios that differ from each other in the antenna load only (see Figure 9.1). The impulsive EM plane wave is defined by its signature $e^i(t)$ and by its polarization and propagation vectors defined via the polar angle θ (see section 7.1). The incident EM plane wave impinges upon the conducting body of the wire antenna, which induces the electric current. The latter subsequently becomes the source of scattered EM wave fields propagating away from the receiving antenna. Accordingly, the far-field EM scattering characteristics of the antenna can be calculated directly, for any antenna load and a direction of observation ξ, from the corresponding induced

---

Time-Domain Electromagnetic Reciprocity in Antenna Modeling, First Edition. Martin Štumpf.

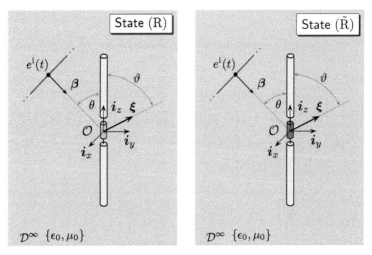

FIGURE 9.1. Receiving situations differing from each other in the antenna load.

electric-current distribution (see section 6.3). Alternatively, one may employ a consequence of EM reciprocity and interrelate the change of EM scattering in states (R) and $(\tilde{R})$ with the transmitting situation depicted in Figure 7.1. With reference to [4, chapter 9], this way is briefly described in the ensuing section. The receiving antennas are placed in the unbounded, homogeneous, loss-free, and isotropic embedding $\mathcal{D}^\infty$.

9.2 PROBLEM SOLUTION

The point of departure for our EM reciprocity analysis is the single-port TD counterpart of the reciprocity relation [4, eq. (9.16)] that is written in terms of far-field amplitudes, that is

$$\Delta \boldsymbol{E}^{s;\infty}(\boldsymbol{\xi},t) *_t V^{\mathrm{T}}(t) = \boldsymbol{E}^{\mathrm{T};\infty}(\boldsymbol{\xi},t) *_t \Delta V^{\mathrm{R}}(t) \qquad (9.1)$$

in which

- $V^{\mathrm{T}}(t)$ = the excitation voltage pulse in the gap of the antenna in state (T);
- $\boldsymbol{E}^{\mathrm{T};\infty}(\boldsymbol{\xi},t)$ = electric-field (vectorial) amplitude radiation characteristics of the transmitting antenna observed at the direction specified by a unit vector of observation $\boldsymbol{\xi}$;
- $\Delta V^{\mathrm{R}}(t) = V^{\tilde{\mathrm{R}}}(t) - V^{\mathrm{R}}(t)$ = the difference of the induced load voltage responses in states $(\tilde{\mathrm{R}})$ and (R);
- $\Delta \boldsymbol{E}^{s;\infty}(\boldsymbol{\xi},t)$ = the change of the electric-field (vectorial) amplitude scattering characteristics of the receiving antenna observed at the direction specified by a unit vector of observation $\boldsymbol{\xi}$.

In terms of the corresponding polar components, we further rewrite Eq. (9.1) to the following form

$$\Delta E_\theta^{s;\infty}(\vartheta, t) *_t V^T(t) = E_\theta^{T;\infty}(\vartheta, t) *_t \Delta V^R(t) \tag{9.2}$$

where ϑ denotes the angle of observation at which the change of EM scattering characteristics is observed (see Figure 9.1). The reciprocity relation (9.2) will next be evaluated numerically. At first, the change of the polar component of the far-field EM scattering characteristics $\Delta E_\theta^{s;\infty}(\vartheta, t)$ is, in virtue of EM reciprocity, calculated from the corresponding radiated far-field amplitude, $E_\theta^{T;\infty}(\vartheta, t)$, and from the change of the load voltage $\Delta V^R(t)$. The latter is found as the difference of the load responses induced by an incident EM plane wave in states $(\tilde{R})$ and (R). Subsequently, $\Delta E_\theta^{s;\infty}(\vartheta, t)$ is found directly from the calculated induced current distributions according to section 6.3.

ILLUSTRATIVE NUMERICAL EXAMPLE

- Make use of the reciprocity relation (9.2) to calculate the change of EM scattering characteristics of a wire due to the change in its lumped load. Subsequently, validate the result by evaluating the impact of the variable load directly from the induced electric-current distributions calculated according to sections 2.3.

Solution: We shall analyze EM scattering from a receiving thin-wire antenna whose length and radius are $\ell = 0.10$ m and $a = 0.10$ mm, respectively. The antenna body extends along the z-axis at $\{r = 0, -0.050$ m $< z < 0.050$ m$\}$ and is at its center loaded by a resistor with $R = 1.0$ kΩ (state (R)) and by a capacitor with $C = 1.0$ pF (state ($\tilde{R}$)). The receiving antenna is in both states ($\tilde{R}$) and (R) irradiated by an impulsive plane wave whose signature has the shape of a bipolar triangle, that is

$$e^i(t) = \frac{2e_m}{t_w} \left[t \, H(t) - 2 \left(t - \frac{t_w}{2} \right) H \left(t - \frac{t_w}{2} \right) \right.$$
$$\left. + 2 \left(t - \frac{3t_w}{2} \right) H \left(t - \frac{3t_w}{2} \right) - (t - 2t_w) H(t - 2t_w) \right] \tag{9.3}$$

where we take $e_m = 1.0$ V/m and $c_0 t_w = 1.0 \, \ell$. The corresponding pulse shape is then similar to the one shown in Figure 7.2b, but with the unit amplitude. The direction of propagation of the incident EM plane wave is specified by $\theta = \pi/4$ (see Figure 9.1). The corresponding induced voltage responses across the chosen lumped loads along with their difference are shown in Figure 9.2.

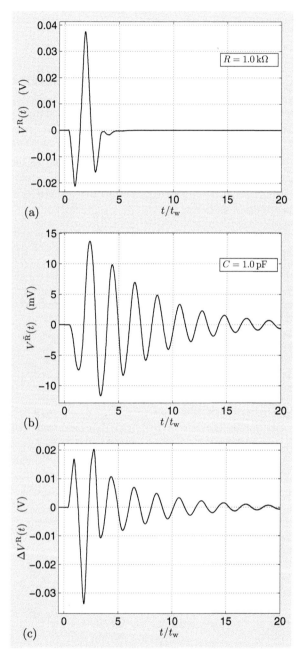

FIGURE 9.2. Load voltage response in (a) state (R) with the resistive load $R = 1.0$ kΩ; (b) state ($\tilde{\text{R}}$) with the capacitive load $C = 1.0$ pF; and (c) load voltage difference.

In the corresponding transmitting situation (T) (see Figure 7.1), the antenna is at its center activated by the voltage pulse that is defined by Eq. (7.4), where we take $V_{\mathrm{m}} = 1.0$ V and $c_0 t_{\mathrm{w}} = 1.0\,\ell$, again (see Figure 7.2a). The resulting electric current distribution is used as the input for calculating the radiated far-field amplitude in a given direction of observation (see section 6.3). In the present numerical example, we take $\vartheta = \pi/8$. The pulse shape of the corresponding far-field radiated amplitude, $E_\theta^{\mathrm{T};\infty}(\vartheta = \pi/8, t)$, is then identical with $E_\theta^{\mathrm{T};\infty}(\theta = \pi/8, t)$ as shown in Figure 7.4. Consequently, with the voltage load difference and the far-field radiated amplitude at our disposal, we may calculate the time convolution on the right-hand side of Eq. (9.2) and get, say, $R(t) = V^{\mathrm{T}}(t) *_t \Delta E_\theta^{\mathrm{s};\infty}(\vartheta, t)$. Again, the desired change of the far-field scattering amplitude follows upon carrying out a deconvolution procedure. An example of the latter can be described by (cf. Eq. (7.5))

$$\Delta E_\theta^{\mathrm{s};\infty}(\vartheta, t) = \frac{t_{\mathrm{w}}^2}{4 V_{\mathrm{m}}} \sum_{n=0}^{\infty} \frac{2(n+2)^2 - 1 + (-1)^{n+2}}{8} \partial_t^3 R(t - n t_{\mathrm{w}}/2) \qquad (9.4)$$

which applies to the excitation voltage pulse $V^{\mathrm{T}}(t)$ defined by Eq. (7.4). The resulting pulse shape calculated with the aid of the reciprocity-based relation (9.2) is shown in Figure 9.3 (see "RECIPROCITY"). For validation purposes, the change of the EM scattering far-field amplitude has also been calculated directly by analyzing ($\tilde{\mathrm{R}}$) and (R) states with the aid of chapters 2 and 6 (see "DIRECT" in Figure 9.3). As can be seen, the obtained pulse shapes are on top of each other, thereby proving the consistency of the proposed modeling methodologies.

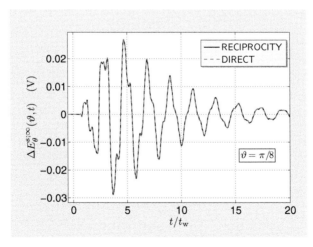

FIGURE 9.3. The change of EM far-field scattering amplitude at $\vartheta = \pi/8$ as calculated via EM reciprocity ("RECIPROCITY") and directly ("DIRECT") from the induced electric-current distribution.

CHAPTER 10

INFLUENCE OF A WIRE SCATTERER ON A RECEIVING WIRE ANTENNA

In chapter 8, we have analyzed the influence of a thin-wire scatterer on the electric-current self-response of a gap-excited transmitting antenna with the help of the EM reciprocity theorem of the time-convolution type. In the present chapter, we shall demonstrate that the reciprocity theorem is also a useful tool for evaluating the impact of such a scatterer on the operation of a receiving antenna. To that end, we shall heavily rely on the conclusions drawn in [4, section 6.1] that are briefly reviewed in section 10.2.

10.1 PROBLEM DESCRIPTION

In this chapter, the effect of a thin-wire PEC scatterer on the induced voltage and electric current across a lumped load of a receiving antenna is evaluated. EM fields in the analyzed problem configurations are excited by one and the same impulsive EM plane wave, so that the receiving situations differ from each other in the presence of the wire scatterer only (see Figure 10.1).

The scattering wire is oriented parallel to the receiving antenna, and their horizontal offset is denoted by $D > 0$, again. Both the antenna and the scatterer are placed in the unbounded, homogeneous, loss-free, and isotropic embedding $\mathcal{D}^\infty$.

10.2 PROBLEM SOLUTION

The starting point for the ensuing analysis is the TD counterpart of the reciprocity relation [4, eq. (6.6)] that is adapted to the problem configuration consisting of a one-port receiving antenna and a thin-wire scatterer. Assuming further the PEC scatterer occupying a bounded domain $\mathcal{L}$ along which the relevant explicit-type boundary condition applies, the reciprocity relation has the following form

Time-Domain Electromagnetic Reciprocity in Antenna Modeling, First Edition. Martin Štumpf.
© 2020 by The Institute of Electrical and Electronics Engineers, Inc. Published 2020 by John Wiley & Sons, Inc.

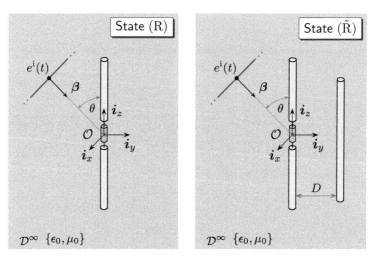

FIGURE 10.1. Receiving scenarios differing from each other in the presence of a wire scatterer.

$$\Delta V^{\mathrm{R}}(t) *_t I^{\mathrm{T}}(t) + V^{\mathrm{T}}(t) *_t \Delta I^{\mathrm{R}}(t)$$

$$= -\int_{\mathcal{L}} E_z^{\mathrm{T}}(D, \zeta, t) *_t I^{\tilde{\mathrm{s}}}(D, \zeta, t)\mathrm{d}\zeta \qquad (10.1)$$

where

- $\Delta V^{\mathrm{R}}(t) = V^{\tilde{\mathrm{R}}}(t) - V^{\mathrm{R}}(t)$ = the difference of the induced load voltage responses in states $(\tilde{\mathrm{R}})$ and (R);
- $\Delta I^{\mathrm{R}}(t) = I^{\tilde{\mathrm{R}}}(t) - I^{\mathrm{R}}(t)$ = the difference of the electric currents flowing across the load in states $(\tilde{\mathrm{R}})$ and (R);
- $V^{\mathrm{T}}(t)$ = the excitation voltage pulse in the gap of the antenna in state (T) (see Figure 7.1);
- $I^{\mathrm{T}}(t)$ = the electric-current response of the transmitting wire antenna at the position of its voltage-gap excitation;
- $E_z^{\mathrm{T}}(D, \zeta, t)$ = the z-component of the total electric field along $\mathcal{L}$ in the *absence* of the wire scatterer. The radiated field can be calculated through formulas (6.12)–(6.14);
- $I^{\tilde{\mathrm{s}}}(D, \zeta, t)$ = the axial electric current induced along the scatterer in state $(\tilde{\mathrm{R}})$.

Since the change of the equivalent Norton's electric-current source strength can be expressed through the input TD admittance of the transmitting antenna, $Y^{\mathrm{T}}(t)$, as

$$\Delta I^{\mathrm{G}}(t) = I^{\tilde{\mathrm{G}}}(t) - I^{\mathrm{G}}(t) = \Delta I^{\mathrm{R}}(t) + Y^{\mathrm{T}}(t) *_t \Delta V^{\mathrm{R}}(t) \qquad (10.2)$$

we may rewrite Eq. (10.1) to its equivalent form (cf. Eq. (8.2) and [4, eq. (6.22)])

$$V^{\mathrm{T}}(t) *_t \Delta I^{\mathrm{G}}(t) = - \int_{\mathcal{L}} E_z^{\mathrm{T}}(D, \zeta, t) *_t I^{\tilde{s}}(D, \zeta, t) \mathrm{d}\zeta \tag{10.3}$$

which is handled in the same way as Eq. (8.2), that is

$$V^{\mathrm{T}}(t) *_t \Delta I^{\mathrm{G}}(t) \simeq -\Delta \sum_{n=1}^{N} E_z^{\mathrm{T}}(D, \zeta_n, t) *_t I^{\tilde{s}}(D, \zeta_n, t) \tag{10.4}$$

where N is the number of discretization nodes along the wire scatterer, and Δ denotes the length of the discretization segments. Finally recall that $I^{\tilde{s}}(D, \zeta_n, t)$ in the reciprocity relation (10.3) is the induced electric current in the (receiving) state $(\tilde{\mathrm{R}})$ (see Figure 10.1), while $I^{\tilde{s}}(D, \zeta_n, t)$ in Eq. (8.2) has the meaning of the induced electric current in the (transmitting) state $(\tilde{\mathrm{T}})$ (see Figure 8.1).

ILLUSTRATIVE NUMERICAL EXAMPLE

- Make use of the reciprocity relation (10.3) to evaluate the impact of a PEC wire scatterer on the TD equivalent Norton's short-circuit electric current of a receiving wire antenna. Subsequently, validate the result by calculating the change of the response directly from the electric-current distributions calculated according to sections 2.3 and 3.3.

Solution: We shall analyze the impact of a PEC thin-wire scatterer placed along $\{r = D = 0.050\,\mathrm{m}, -0.075\,\mathrm{m} < z < 0.075\,\mathrm{m}\}$ on the receiving antenna extending along the z-axis at $\{r = 0, -0.050\,\mathrm{m} < z < 0.050\,\mathrm{m} = \ell/2\}$. Again, both wire structures have a radius $a = 0.10\,\mathrm{mm}$. The receiving antenna is at its center short-circuited, so that $V^{\mathrm{R}}(t) = V^{\tilde{\mathrm{R}}}(t) = \Delta V^{\mathrm{R}}(t) = 0$. The EM fields in states (R) and $(\tilde{\mathrm{R}})$ are excited via an incident EM plane wave whose signature is defined by Eq. (9.3) with $e_{\mathrm{m}} = 1.0\,\mathrm{V/m}$ and $c_0 t_{\mathrm{w}} = 1.0\,\ell$. The corresponding pulse shape is then similar to the one shown in Figure 7.2b, but with the unit amplitude.

In the first step, the change of the Norton current is found with the aid of EM reciprocity. To that end, we evaluate the right-hand side of Eq. (10.4). As the radiated electric-field pulses $E_z^{\mathrm{T}}(D, \zeta_n, t)$ in absence of the scatterer have been calculated before (see Figure 8.2a), the remaining task is to find the induced electric-current distribution along the scatterer in state $(\tilde{\mathrm{R}})$. This step can be accomplished with the help of approaches described in sections 2.3 with 2.4.1 and 3.3. Examples of the corresponding pulse shapes are shown in Figure 10.2a. For the sake of comparison, we have also plotted the corresponding electric current pulses along the scatterer in absence of the receiving antenna (see Figure 10.2b). Despite the relatively small

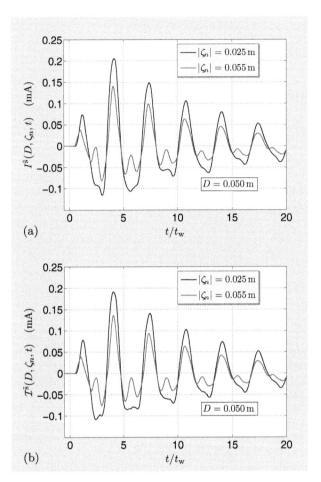

FIGURE 10.2. Induced electric-current pulses on the scatterer. (a) In state $(\tilde{R})$; (b) in absence of the receiving antenna.

horizontal offset between the wires, that is $D = c_0 t_w/2$, the corresponding pulse shapes do not differ significantly.

Next, with the set of the radiated electric-field and the induced electric-current pulses at our disposal, we can evaluate the right-hand side of Eq. (10.4) and get $P(t) = V^T(t) *_t \Delta I^G(t)$. Subsequently, the change of the Norton current follows through deconvolution (cf. Eq. (8.4))

$$\Delta I^G(t) = \frac{t_w^2}{4V_m} \sum_{n=0}^{\infty} \frac{2(n+2)^2 - 1 + (-1)^{n+2}}{8} \partial_t^3 P(t - nt_w/2) \qquad (10.5)$$

which gives the pulse shape shown in Figure 10.3b ("RECIPROCITY"). For validation purposes, the electric-current change has also been calculated directly

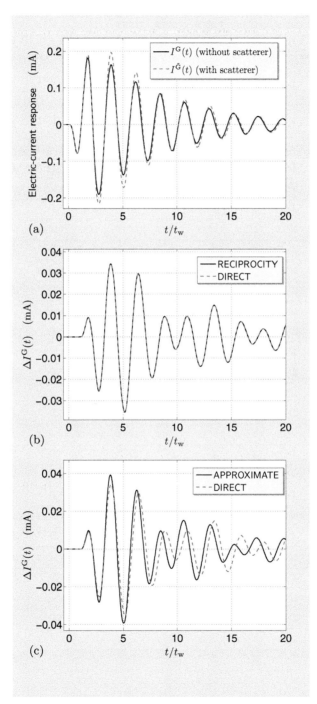

FIGURE 10.3. (a) Norton's electric-current response of the antenna with and without the scatterer. The change of the response as evaluated (b) directly and using EM reciprocity and (c) approximately.

by analyzing the corresponding receiving states $(\tilde{\mathrm{R}})$ and (R) (see "DIRECT" in Figure 10.3b). As can be seen, the results plotted in Figure 10.3b are hardly distinguishable, which validates the modeling methodologies described in chapters 2, 3, and 6.

Finally, the change of the Norton current has been evaluated approximately by replacing $I^{\tilde{s}}(D, \zeta_n, t)$ in Eq. (10.4) with the electric-current distribution $\mathcal{I}^{\tilde{s}}(D, \zeta_n, t)$ calculated in absence of the receiving antenna (see Figure 10.2). This way avoids calculating the double integrations in Eq. (D.15), which significantly accelerates the numerical solution procedure. From Figure 10.3c, it is clear, however, that this approximation is appropriate for the early-time part of the response only.

CHAPTER 11

EM-FIELD COUPLING TO TRANSMISSION LINES

In this chapter, the EM reciprocity theorem of the time-convolution type is systematically applied to introduce an EM-field-to-line coupling model enabling to calculate the induced voltage and current quantities at the ends of a transmission line through testing voltage/current quantities pertaining to the situation when the line operates as a transmitter. Subsequently, a relation of the reciprocity-based coupling model to the classic "scattered-voltage" model due to Agrawal et al. [34] is demonstrated. Finally, EM reciprocity is employed again to show that the EM wave fields radiated in the testing state can replace the excitation EM-field distribution along the line, thus providing yet alternative coupling models for both an EM plane-wave incidence and a known EM source distribution.

11.1 INTRODUCTION

To reveal some aspects of the transmission-line model, we shall first analyze EM field radiation from a PEC wire of radius $a > 0$ that is located above a PEC ground in the homogeneous, isotropic, and loss-free half-space $\mathcal{D}_+^\infty$ (see Figure 11.1). EM properties of the latter are described by (real-valued, positive and scalar) electric permittivity ϵ_0 and magnetic permeability μ_0, which implies the EM wave speed $c_0 = (\epsilon_0\mu_0)^{-1/2} > 0$. The wire is excited via the voltage pulse $V^T(t)$ that is applied in a narrow gap located around $x = 0$. The excitation pulse induces an axial electric-current distribution along the wire that subsequently radiates EM fields into the half-space.

If the gap-excited wire is sufficiently long such that the effect of the wire's ends can be neglected, the axial component of the scattered electric-field strength can be represented as (cf. Eqs. (B.5) and (B.6))

Time-Domain Electromagnetic Reciprocity in Antenna Modeling, First Edition. Martin Štumpf.
© 2020 by The Institute of Electrical and Electronics Engineers, Inc. Published 2020 by John Wiley & Sons, Inc.

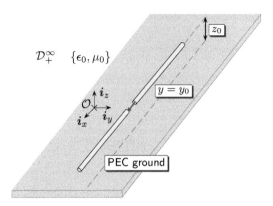

FIGURE 11.1. A gap-excited wire over a PEC ground.

$$\hat{E}_x^{\mathrm{s}}(\boldsymbol{x}, s) = -[s\hat{V}^{\mathrm{T}}(s)/2\pi\mathrm{i}]$$

$$\int_{p=-\mathrm{i}\infty}^{\mathrm{i}\infty} \exp(-spx) \frac{\mathrm{K}_0[s\gamma_0(p)r_0] - \mathrm{K}_0[s\gamma_0(p)r_1]}{\mathrm{K}_0[s\gamma_0(p)a] - \mathrm{K}_0[s\gamma_0(p)2z_0]} \mathrm{d}p \qquad (11.1)$$

where $\{s \in \mathbb{R}; s > 0\}$ is the Laplace-transform parameter, $\gamma_0(p) = (1/c_0^2 - p^2)^{1/2}$ with $\mathrm{Re}[\gamma_0(p)] > 0$ denotes the wave slowness parameter in the radial direction, and $r_0 = [(y - y_0)^2 + (z - z_0)^2]^{1/2}$ with $r_1 = [(y - y_0)^2 + (z + z_0)^2]^{1/2}$ are the distances from the wire and its image, respectively. Apparently, along the wire as $r_0 \downarrow a$, Eq. (11.1) boils down to $\hat{E}_x^{\mathrm{s}} + \hat{V}^{\mathrm{T}}(s)\delta(x) = 0$, while $\hat{E}_x^{\mathrm{s}}(x, y, 0, s) = 0$ for all $(x, y) \in \mathbb{R}^2$ on the ground plane, where $r_0 = r_1$.

The axial electric current along the wire then follows from the azimuthal magnetic-field strength according to (cf. Eq. (B.9))

$$\hat{I}(x, s) = 2\pi a \lim_{r_0 \downarrow a} \hat{H}_\phi^{\mathrm{s}}(\boldsymbol{x}, s)$$

$$= -\mathrm{i}as\epsilon_0 \hat{V}^{\mathrm{T}}(s)$$

$$\times \int_{p=-\mathrm{i}\infty}^{\mathrm{i}\infty} \exp(-spx) \frac{\mathrm{K}_1[s\gamma_0(p)a]}{\mathrm{K}_0[s\gamma_0(p)a] - \mathrm{K}_0[s\gamma_0(p)2z_0]} \frac{\mathrm{d}p}{\gamma_0(p)} \qquad (11.2)$$

The integrand in Eq. (11.2) is a multivalued function of the axial slowness parameter p. An exact evaluation of the integral hence requires to account for the presence of the branch cuts along $\{1/c_0 < |\mathrm{Re}(p)| < \infty, \mathrm{Im}(p) = 0\}$ due to the square root in $\gamma_0(p)$ (see Figure C.1). Under a "low-frequency approximation," however, the small-argument expansions for the modified Bessel functions apply [25, p. 375], and we get

$$\frac{\mathrm{K}_1[s\gamma_0(p)a]}{\mathrm{K}_0[s\gamma_0(p)a] - \mathrm{K}_0[s\gamma_0(p)2z_0]} = \frac{1}{s\gamma_0(p)a} \frac{1}{\ln(2z_0/a)} + O(s) \qquad (11.3)$$

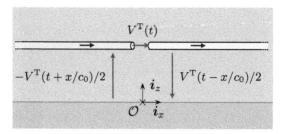

FIGURE 11.2. A gap-excited transmission line over a PEC ground.

as $s \downarrow 0$. Substituting Eq. (11.3) in the expression for the induced current (11.2), we hence arrive at its "low-frequency" approximation

$$\hat{I}(x,s) \simeq \frac{\epsilon_0 \hat{V}^{\mathrm{T}}(s)}{\mathrm{i}\ln(2z_0/a)} \int_{p=-\mathrm{i}\infty}^{\mathrm{i}\infty} \exp(-spx) \frac{\mathrm{d}p}{1/c_0^2 - p^2} \tag{11.4}$$

that can be readily cast into its equivalent form, that is

$$\hat{I}(x,s) \simeq \frac{s\hat{\eta}(s)\hat{V}^{\mathrm{T}}(s)}{2\mathrm{i}\pi} \int_{p=-\mathrm{i}\infty}^{\mathrm{i}\infty} \exp(-spx) \frac{\mathrm{d}p}{\hat{\zeta}\hat{\eta} - s^2 p^2} \tag{11.5}$$

in which

$$\hat{\eta}(s) = 2\pi s\epsilon_0 / \ln(2z_0/a) \tag{11.6}$$

$$\hat{\zeta}(s) = s\mu_0 \ln(2z_0/a)/2\pi \tag{11.7}$$

denote the transverse admittance and the longitudinal impedance (per-unit-length) of the transmission line, respectively. The integrand in the approximate expression (11.5) is no longer a multivalued function of p, but it shows two simple pole singularities at $p = \pm 1/c_0$ only. Their contribution can be interpreted as a pair of plane waves propagating away from the gap source in opposite directions along the wire axis (see Figure 11.2). Hence, evaluating the pole contributions, we get

$$\hat{I}(x,s) \simeq Y^c \hat{V}^{\mathrm{T}}(s) \exp(-s|x|/c_0)/2 \tag{11.8}$$

with

$$Y^c = (\hat{\eta}/\hat{\zeta})^{1/2} = 2\pi\eta_0 / \ln(2z_0/a) \tag{11.9}$$

denoting the characteristic admittance of the transmission line above the PEC perfect ground. It is noted now that this solution could also be found directly from the corresponding transmission-line equations describing a one-dimensional wave motion along a gap-excited transmission line, that is

$$\partial_x \hat{V}(x, s) + \hat{\zeta}(s)\hat{I}(x, s) = \hat{V}^{\mathrm{T}}(s)\delta(x) \qquad (11.10)$$

$$\partial_x \hat{I}(x, s) + \hat{\eta}(s)\hat{V}(x, s) = 0 \qquad (11.11)$$

that are supplied with the boundedness ("radiation") conditions $\{\hat{V}, \hat{I}\}(x, s) = \{0, 0\}$ as $|x| \to \infty$ for $\mathrm{Re}(s) > 0$, thereby accounting for the property of causality.

In conclusion, we have just demonstrated that the transmission-line theory furnishes a "low-frequency approximation" to the (exact) solution satisfying the EM field equations. With this conclusion in mind, we shall next describe EM-field coupling models facilitating the efficient calculation of induced load voltages on transmission lines in the presence of an external EM field disturbance.

11.2 PROBLEM DESCRIPTION

The main objective of the present chapter is to describe the interaction of an external EM field disturbance with a transmission line above a PEC ground. In particular, we are interested in the voltage response at the transmission-line terminals that is induced by an external EM-field disturbance represented by the incident EM wave fields $\{\hat{\boldsymbol{E}}^{\mathrm{i}}, \hat{\boldsymbol{H}}^{\mathrm{i}}\} = \{\hat{\boldsymbol{E}}^{\mathrm{i}}, \hat{\boldsymbol{H}}^{\mathrm{i}}\}(x, s)$ (see Figure 11.3). To that end, we first define the scattered field as the difference between the actual field in the configuration and the excitation field describing the total field if the line were absent, that is

$$\{\hat{\boldsymbol{E}}^{\mathrm{s}}, \hat{\boldsymbol{H}}^{\mathrm{s}}\}(x, s) \triangleq \{\hat{\boldsymbol{E}}^{\mathrm{R}}, \hat{\boldsymbol{H}}^{\mathrm{R}}\}(x, s) - \{\hat{\boldsymbol{E}}^{\mathrm{e}}, \hat{\boldsymbol{H}}^{\mathrm{e}}\}(x, s) \qquad (11.12)$$

This implies that the excitation field is, in fact, a superposition of the incident EM field and the field reflected from the PEC ground. In the ensuing analysis, the reciprocity theorem of the time-convolution type (see [3, section 28.4] and [4, section 1.4.1]) is applied to find an interaction quantity from which the induced voltage at the transmission-line terminals can be obtained.

11.3 EM-FIELD-TO-LINE INTERACTION

The actual (receiving) state defined in the previous section will be next interrelated with the testing state, denoted by (T), in which the transmission line is excited by a lumped source (see Figure 11.4, for example). The desired reciprocity relation is derived in two steps that follow. In the first one, the reciprocity theorem is applied to the domain externally bounded by $\mathcal{S}_0$ and to the total fields in (R) and (T) states (see Table 11.1). Thanks to the explicit-type boundary conditions on the PEC ground plane and the PEC conductor, the surface integral over $\mathcal{S}_0$ can be written as a sum of integrals over closed surfaces surrounding the terminals of the transmission-line

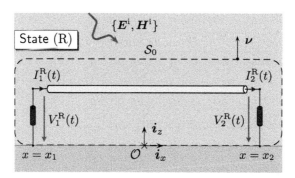

FIGURE 11.3. Actual (receiving) situation (R) in which the transmission line is irradiated by an external EM-field disturbance.

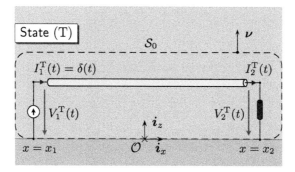

FIGURE 11.4. Testing (transmitting) situation (T) in which the transmission line is activated by a lumped source.

TABLE 11.1. Application of the Reciprocity Theorem

| Time-Convolution | Domain Interior to $\mathcal{S}_0$ | |
	State (R)	State (T)
Source	0	0
Field	$\{\hat{E}^{\mathrm{R}}, \hat{H}^{\mathrm{R}}\}$	$\{\hat{E}^{\mathrm{T}}, \hat{H}^{\mathrm{T}}\}$
Material	$\{\epsilon_0, \mu_0\}$	$\{\epsilon_0, \mu_0\}$

system. Applying the local "low-frequency approximations" to the EM fields in the neighborhood of terminals, the EM field interaction on the terminal surfaces can be expressed in terms of Kirchhoff-network quantities [4, section 1.5.2]. Consequently,

TABLE 11.2. Application to the Reciprocity Theorem

Domain Exterior to $\mathcal{S}_0$		
Time-Convolution	State (s)	State (T)
Source	0	0
Field	$\{\hat{\boldsymbol{E}}^{\mathrm{s}}, \hat{\boldsymbol{H}}^{\mathrm{s}}\}$	$\{\hat{\boldsymbol{E}}^{\mathrm{T}}, \hat{\boldsymbol{H}}^{\mathrm{T}}\}$
Material	$\{\epsilon_0, \mu_0\}$	$\{\epsilon_0, \mu_0\}$

denoting the voltages and currents at the end points at $x = x_{1,2}$ by $\{\hat{V}_{1,2}, \hat{I}_{1,2}\}$, respectively, we end up with

$$\int_{\boldsymbol{x} \in \mathcal{S}_0} \left(\hat{\boldsymbol{E}}^{\mathrm{R}} \times \hat{\boldsymbol{H}}^{\mathrm{T}} - \hat{\boldsymbol{E}}^{\mathrm{T}} \times \hat{\boldsymbol{H}}^{\mathrm{R}} \right) \cdot \nu \mathrm{d}A$$
$$\simeq \hat{V}_1^{\mathrm{R}}(s)\hat{I}_1^{\mathrm{T}}(s) - \hat{V}_1^{\mathrm{T}}(s)\hat{I}_1^{\mathrm{R}}(s) - \hat{V}_2^{\mathrm{R}}(s)\hat{I}_2^{\mathrm{T}}(s) + \hat{V}_2^{\mathrm{T}}(s)\hat{I}_2^{\mathrm{R}}(s) \qquad (11.13)$$

where we have taken into account the orientation of electric currents with respect to the outer unit vector on the surfaces bounding the terminations and the self-adjointness of the states. Secondly, the reciprocity theorem is applied to the scattered (s) and transmitting (T) field states and to the unbounded domain exterior to $\mathcal{S}_0$ according to Table 11.2. With the aid of the conclusions applying to the reciprocity theorem of the time-convolution type on unbounded domains (see [4, section 1.4.3]) and with the explicit-type boundary conditions on the PEC ground plane in mind, we arrive at

$$\int_{\boldsymbol{x} \in \mathcal{S}_0} (\hat{\boldsymbol{E}}^{\mathrm{s}} \times \hat{\boldsymbol{H}}^{\mathrm{T}} - \hat{\boldsymbol{E}}^{\mathrm{T}} \times \hat{\boldsymbol{H}}^{\mathrm{s}}) \cdot \nu \mathrm{d}A = 0 \qquad (11.14)$$

Finally, upon combining Eqs. (11.13)–(11.14) with (11.12), we get

$$\hat{V}_1^{\mathrm{R}}(s)\hat{I}_1^{\mathrm{T}}(s) - \hat{V}_1^{\mathrm{T}}(s)\hat{I}_1^{\mathrm{R}}(s) - \hat{V}_2^{\mathrm{R}}(s)\hat{I}_2^{\mathrm{T}}(s) + \hat{V}_2^{\mathrm{T}}(s)\hat{I}_2^{\mathrm{R}}(s)$$
$$\simeq \int_{\boldsymbol{x} \in \mathcal{S}_0} (\hat{\boldsymbol{E}}^{\mathrm{e}} \times \hat{\boldsymbol{H}}^{\mathrm{T}} - \hat{\boldsymbol{E}}^{\mathrm{T}} \times \hat{\boldsymbol{H}}^{\mathrm{e}}) \cdot \nu \mathrm{d}A \qquad (11.15)$$

whose right-hand side can be further developed with the aid of the explicit-type boundary conditions applying on the PEC ground plane and the PEC conductor. In this way, we find

$$\int_{\boldsymbol{x} \in \mathcal{S}_0} (\hat{\boldsymbol{E}}^{\mathrm{e}} \times \hat{\boldsymbol{H}}^{\mathrm{T}} - \hat{\boldsymbol{E}}^{\mathrm{T}} \times \hat{\boldsymbol{H}}^{\mathrm{e}}) \cdot \nu \mathrm{d}A = -\int_{\boldsymbol{x} \in \mathcal{S}} \hat{\boldsymbol{E}}^{\mathrm{e}} \cdot \partial \hat{\boldsymbol{J}}^{\mathrm{T}} \, \mathrm{d}A \qquad (11.16)$$

where $\partial \hat{\boldsymbol{J}}^{\mathrm{T}}(x, s) \triangleq \nu(x) \times \hat{\boldsymbol{H}}^{\mathrm{T}}(x, s)$ denotes the testing electric-current surface density on the cylindrical surface $\mathcal{S}$ enclosing the transmission line. Assuming that the testing current is essentially concentrated along the transmission line's axis (including its vertical sections), we finally arrive at

$$\hat{V}_1^{\mathrm{R}}(s)\hat{I}_1^{\mathrm{T}}(s) - \hat{V}_1^{\mathrm{T}}(s)\hat{I}_1^{\mathrm{R}}(s) - \hat{V}_2^{\mathrm{R}}(s)\hat{I}_2^{\mathrm{T}}(s) + \hat{V}_2^{\mathrm{T}}(s)\hat{I}_2^{\mathrm{R}}(s)$$

$$\simeq -\int_{x=x_1}^{x_2} \hat{E}_x^{\mathrm{e}}(x,y_0,z_0,s)\hat{I}^{\mathrm{T}}(x,s)\mathrm{d}x$$

$$- \hat{I}_1^{\mathrm{T}}(s)\int_{z=0}^{z_0} \hat{E}_z^{\mathrm{e}}(x_1,y_0,z,s)\mathrm{d}z + \hat{I}_2^{\mathrm{T}}(s)\int_{z=0}^{z_0} \hat{E}_z^{\mathrm{e}}(x_2,y_0,z,s)\mathrm{d}z \qquad (11.17)$$

The first term on the right-hand side of Eq. (11.17) accounts for the excitation-field distribution along the horizontal section of the transmission line extending over $\{x_1 < x < x_2, y = y_0, z = z_0 > 0\}$, while the remaining terms describe the contributions from the (vertical) terminal sections along $\{x = x_{1,2}, y = y_0, 0 < z < z_0\}$. As the reciprocity relation (11.17) interrelates the terminal voltage and current quantities with the distribution of the excitation field along the transmission line, it can be viewed as a generalization of the so-called "Baum-Liu-Tesche equations" (see, e.g. [35, section 7.2.4]). For an illustrative application of the reciprocity-based coupling model to the calculation of lightning-induced voltages on a transmission line, we refer the reader to [36].

11.4   RELATION TO AGRAWAL COUPLING MODEL

The most popular EM-field-to-line coupling model was introduced by Agrawal et al. [34]. In this section, we shall reveal its relation to the reciprocity-based model described by Eq. (11.17). From this reason, we shall first define the "excitation voltage"

$$\hat{V}^{\mathrm{e}}(x,s) \triangleq -\int_{z=0}^{z_0} \hat{E}_z^{\mathrm{e}}(x,y_0,z,s)\mathrm{d}z \qquad (11.18)$$

that makes it possible to rewrite Eq. (11.17) in the following form

$$\int_{x=x_1}^{x_2} \partial_x[\hat{V}^{\mathrm{T}}(x,s)\hat{I}^{\mathrm{R}}(x,s) - \hat{V}^{\mathrm{R}}(x,s)\hat{I}^{\mathrm{T}}(x,s)]\mathrm{d}x$$

$$\simeq -\int_{x=x_1}^{x_2} \hat{E}_x^{\mathrm{e}}(x,y_0,z_0,s)\hat{I}^{\mathrm{T}}(x,s)\mathrm{d}x$$

$$- \int_{x=x_1}^{x_2} \partial_x[\hat{I}^{\mathrm{T}}(x,s)\hat{V}^{\mathrm{e}}(x,s)]\mathrm{d}x \qquad (11.19)$$

from which we conclude that

$$\partial_x[\hat{V}^{\mathrm{T}}(x,s)\hat{I}^{\mathrm{R}}(x,s) - \hat{V}^{\mathrm{R}}(x,s)\hat{I}^{\mathrm{T}}(x,s)]$$

$$\simeq -\hat{E}_x^{\mathrm{e}}(x,y_0,z_0,s)\hat{I}^{\mathrm{T}}(x,s) - \partial_x[\hat{I}^{\mathrm{T}}(x,s)\hat{V}^{\mathrm{e}}(x,s)] \qquad (11.20)$$

provided that the expressions under the integral signs are continuous functions with respect to x throughout $\{x_1 < x < x_2\}$. Defining now the "scattered voltage," that is $\hat{V}^s \triangleq \hat{V}^R - \hat{V}^e$ (cf. Eq. (11.12)), we finally arrive at the desired interaction quantity of the local-type

$$\partial_x \left[\hat{V}^T(x,s)\hat{I}^R(x,s) - \hat{V}^s(x,s)\hat{I}^T(x,s) \right]$$
$$\simeq -\hat{E}_x^e(x,y_0,z_0,s)\hat{I}^T(x,s) \tag{11.21}$$

for all $x \in (x_1, x_2)$. Now it remains to show that Eq. (11.21) can also be derived by interrelating the following pair of transmission-line equations (cf. [37, eqs. (18)–(19)])

$$\partial_x \hat{V}^s(x,s) + \hat{\zeta}(x,s)\hat{I}^R(x,s) = \hat{E}_x^e(x,y_0,z_0,s) \tag{11.22}$$
$$\partial_x \hat{I}^R(x,s) + \hat{\eta}(x,s)\hat{V}^s(x,s) = 0 \tag{11.23}$$

with the one applying to testing voltage and current quantities, that is

$$\partial_x \hat{V}^T(x,s) + \hat{\zeta}(x,s)\hat{I}^T(x,s) = 0 \tag{11.24}$$
$$\partial_x \hat{I}^T(x,s) + \hat{\eta}(x,s)\hat{V}^T(x,s) = 0 \tag{11.25}$$

with $\hat{\zeta}(x,s) = sL(x)$ and $\hat{\eta}(x,s) = sC(x)$ applying to a loss-free transmission line with the (per-unit-length) inductance $L(x)$ and capacitance $C(x)$. To demonstrate the relation, Eq. (11.22) multiplied by $\hat{I}^T$ is first added to Eq. (11.25) multiplied by $\hat{V}^s$, and the resulting expression is subsequently subtracted from the expression found by adding Eq. (11.24) multiplied by $\hat{I}^R$ to Eq. (11.23) multiplied by $\hat{V}^T$.

The system of Eqs. (11.22) and (11.23) are in the literature referred to as Agrawal's model. If these equations are chosen as the starting point for solving the coupling problem, then the corresponding boundary conditions must be specified to fully define the boundary-value problem, that is

$$\hat{V}_1^s(s) + \hat{V}_1^e(s) = -\hat{Z}_1^L(s)\hat{I}_1^R(s) \tag{11.26}$$
$$\hat{V}_2^s(s) + \hat{V}_2^e(s) = \hat{Z}_2^L(s)\hat{I}_2^R(s) \tag{11.27}$$

where $\hat{Z}_{1,2}^L(s)$ denotes load impedances at $x_{1,2}$, respectively, and $\hat{V}_{1,2}^e(s)$ are the excitation voltages at $x_{1,2}$, respectively, that directly follow from Eq. (11.18). Finally note that the local-type interaction quantity (11.21) and hence the equations of Agrawal have been derived under the assumption that $\{\hat{V}^s, \hat{I}^R\}(x,s)$ are continuously differentiable functions with respect to x throughout $\{x_1 < x < x_2\}$, which applies to a transmission line whose per-unit-length impedance and admittance vary continuously with x in the interval. The reciprocity-based formulation given by Eq. (11.17) is not limited in this sense and encompasses transmission lines whose per-unit-length parameters are piecewise continuous functions in x. Finally note that the link to other couplings models (see [38, 39], for example) can be revealed along the same lines. This task is left as an exercise for the reader (see [40]).

11.5 ALTERNATIVE COUPLING MODELS BASED ON EM RECIPROCITY

The EM-reciprocity coupling model (11.17) calls for the evaluation of the excitation electric-field distribution along the transmission line. In the present section, EM reciprocity is employed to show that the excitation-field distribution can be replaced with EM wave fields that are radiated from the transmission line in the testing situation. As in [28, sections VI and VII] and [4, section 5.1], the results are provided for both the plane-wave incidence and a known EM-volume-source distribution.

11.5.1 EM Plane-Wave Incidence

We assume that the transmission line is in its (actual) receiving state (R) irradiated by a uniform EM plane wave defined by

$$\hat{\boldsymbol{E}}^{\mathrm{i}}(\boldsymbol{x}, s) = \alpha \hat{e}^{\mathrm{i}}(s) \exp(-s\boldsymbol{\beta} \cdot \boldsymbol{x}/c_0) \tag{11.28}$$

where $\boldsymbol{\alpha}$ denotes the polarization (unit) vector, $\boldsymbol{\beta}$ is a unit vector in the direction of propagation, and $\hat{e}^{\mathrm{i}}(s)$ is the plane-wave amplitude. The excitation EM wave field that accounts for the presence of the PEC ground plane through the reflected wave (see Eq. (11.12)) is hence written as

$$\hat{\boldsymbol{E}}^{\mathrm{e}}(\boldsymbol{x}, s) = (\boldsymbol{\alpha}_{\parallel} + \boldsymbol{\alpha}_{\perp})\hat{e}^{\mathrm{i}}(s) \exp[-s(\boldsymbol{\beta}_{\parallel} + \boldsymbol{\beta}_{\perp}) \cdot \boldsymbol{x}/c_0]$$
$$- (\boldsymbol{\alpha}_{\parallel} - \boldsymbol{\alpha}_{\perp})\hat{e}^{\mathrm{i}}(s) \exp[-s(\boldsymbol{\beta}_{\parallel} - \boldsymbol{\beta}_{\perp}) \cdot \boldsymbol{x}/c_0] \tag{11.29}$$

where the polarization and propagation vectors have been decomposed into their components parallel ($\parallel$) and perpendicular ($\perp$) with respect to the ground plane. Consequently, it is seen that $\boldsymbol{\alpha}_{\parallel} \cdot \hat{\boldsymbol{E}}^{\mathrm{e}} = 0$ on the PEC ground plane, where $\boldsymbol{\beta}_{\perp} \cdot \boldsymbol{x} = 0$. The surface-integral representation of the radiated EM field in the testing state is given as (cf. [4, eq. (4.11)])

$$\hat{\boldsymbol{E}}^{\mathrm{T};\infty}(\boldsymbol{\xi}, s) = s\mu_0(\boldsymbol{\xi}\boldsymbol{\xi} - \boldsymbol{I}) \cdot \int_{\boldsymbol{x} \in \mathcal{S}_0} \exp(s\boldsymbol{\xi} \cdot \boldsymbol{x}/c_0)[\boldsymbol{\nu}(\boldsymbol{x}) \times \hat{\boldsymbol{H}}^{\mathrm{T}}(\boldsymbol{x}, s)]\mathrm{d}A$$
$$+ sc_0^{-1}\boldsymbol{\xi} \times \int_{\boldsymbol{x} \in \mathcal{S}_0} \exp(s\boldsymbol{\xi} \cdot \boldsymbol{x}/c_0)[\hat{\boldsymbol{E}}^{\mathrm{T}}(\boldsymbol{x}, s) \times \boldsymbol{\nu}(\boldsymbol{x})]\mathrm{d}A \tag{11.30}$$

where $\boldsymbol{I}$ denotes the identity tensor. The radiation from the electric-current surface distribution induced on the PEC ground plane can be accounted for by the corresponding image source (see Figure 11.5), and we may hence write (cf. Eq. (11.16))

$$\int_{\boldsymbol{x} \in \mathcal{S}_0} (\hat{\boldsymbol{E}}^{\mathrm{e}} \times \hat{\boldsymbol{H}}^{\mathrm{T}} - \hat{\boldsymbol{E}}^{\mathrm{T}} \times \hat{\boldsymbol{H}}^{\mathrm{e}}) \cdot \boldsymbol{\nu}\mathrm{d}A$$
$$= \hat{e}^{\mathrm{i}}(s)\boldsymbol{\alpha} \cdot \hat{\boldsymbol{E}}^{\mathrm{T};\infty}(-\boldsymbol{\beta}, s)/s\mu_0 \tag{11.31}$$

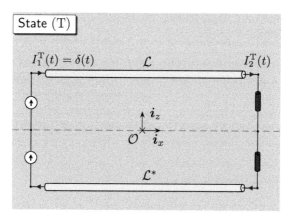

FIGURE 11.5. Transmission line in state (T) and its image to account for the presence of PEC ground plane.

upon combining Eqs. (11.29) and (11.30). Consequently, adding the latter relation to Eq. (11.14) and using (11.12) with (11.13), we end up with

$$\hat{V}_1^{\mathrm{R}}(s)\hat{I}_1^{\mathrm{T}}(s) - \hat{V}_1^{\mathrm{T}}(s)\hat{I}_1^{\mathrm{R}}(s) - \hat{V}_2^{\mathrm{R}}(s)\hat{I}_2^{\mathrm{T}}(s) + \hat{V}_2^{\mathrm{T}}(s)\hat{I}_2^{\mathrm{R}}(s)$$

$$\simeq \hat{e}^{\mathrm{i}}(s)\,\boldsymbol{\alpha} \cdot \hat{\boldsymbol{E}}^{\mathrm{T};\infty}(-\boldsymbol{\beta}, s)/s\mu_0 \tag{11.32}$$

To conclude, we have just demonstrated that the right-hand side of the reciprocity relation (11.17) can be for the plane-wave incidence expressed using the (co-polarized component of the) far-field radiated amplitude of the electric-type as observed in the backward direction, that is, in the opposite direction of propagation of the incident plane wave (see Eq. (11.28)).

11.5.2   Known EM Source Distribution

We shall now assume that the transmission line is in its (actual) receiving situation (see Figure 11.3) irradiated by an incident EM wave field generated by the known source distribution described via electric- and magnetic-current volume densities, $\{\hat{\boldsymbol{J}}^{\mathrm{R}}, \hat{\boldsymbol{K}}^{\mathrm{R}}\} = \{\hat{\boldsymbol{J}}^{\mathrm{R}}, \hat{\boldsymbol{K}}^{\mathrm{R}}\}(\boldsymbol{x}, s)$, respectively, of a bounded support $\mathcal{D}^S$ located exterior to $\mathcal{S}_0$. Consequently, application of the reciprocity theorem of the time-convolution type to the (unbounded) domain exterior to $\mathcal{S}_0$ and to the total fields in states (R) and (T) according to Table 11.3 yields

$$\int_{\boldsymbol{x}\in\mathcal{S}_0}(\hat{\boldsymbol{E}}^{\mathrm{R}} \times \hat{\boldsymbol{H}}^{\mathrm{T}} - \hat{\boldsymbol{E}}^{\mathrm{T}} \times \hat{\boldsymbol{H}}^{\mathrm{R}}) \cdot \boldsymbol{\nu}\mathrm{d}A$$

$$= \int_{\boldsymbol{x}\in\mathcal{D}^S}(\hat{\boldsymbol{K}}^{\mathrm{R}} \cdot \hat{\boldsymbol{H}}^{\mathrm{T}} - \hat{\boldsymbol{J}}^{\mathrm{R}} \cdot \hat{\boldsymbol{E}}^{\mathrm{T}})\,\mathrm{d}V \tag{11.33}$$

TABLE 11.3. Application of the Reciprocity Theorem

Domain Exterior to S_0		
Time-Convolution	State (R)	State (T)
Source	$\{\hat{\boldsymbol{J}}^{\mathrm{R}}, \hat{\boldsymbol{K}}^{\mathrm{R}}\}$	0
Field	$\{\hat{\boldsymbol{E}}^{\mathrm{R}}, \hat{\boldsymbol{H}}^{\mathrm{R}}\}$	$\{\hat{\boldsymbol{E}}^{\mathrm{T}}, \hat{\boldsymbol{H}}^{\mathrm{T}}\}$
Material	$\{\epsilon_0, \mu_0\}$	$\{\epsilon_0, \mu_0\}$

where we have taken into the account the orientation of ν over S_0. Combination of the latter with Eq. (11.13) then leads to

$$\hat{V}_1^{\mathrm{R}}(s)\hat{I}_1^{\mathrm{T}}(s) - \hat{V}_1^{\mathrm{T}}(s)\hat{I}_1^{\mathrm{R}}(s) - \hat{V}_2^{\mathrm{R}}(s)\hat{I}_2^{\mathrm{T}}(s) + \hat{V}_2^{\mathrm{T}}(s)\hat{I}_2^{\mathrm{R}}(s)$$

$$\simeq \int_{\boldsymbol{x} \in \mathcal{D}^S} (\hat{\boldsymbol{K}}^{\mathrm{R}} \cdot \hat{\boldsymbol{H}}^{\mathrm{T}} - \hat{\boldsymbol{J}}^{\mathrm{R}} \cdot \hat{\boldsymbol{E}}^{\mathrm{T}}) \, \mathrm{d}V \qquad (11.34)$$

which is the sought relation. For example, let us take $\hat{\boldsymbol{K}}^{\mathrm{R}} = \boldsymbol{0}$ and assume the field generated by an electric dipole oriented along $\boldsymbol{\ell}^{\mathrm{R}}$ that is specified by

$$\hat{\boldsymbol{J}}^{\mathrm{R}}(\boldsymbol{x}, s) = \hat{\imath}^{\mathrm{R}}(s)\boldsymbol{\ell}^{\mathrm{R}}\delta(\boldsymbol{x} - \boldsymbol{x}^S) \qquad (11.35)$$

where $\delta(\boldsymbol{x} - \boldsymbol{x}^S)$ denotes the three-dimensional Dirac-delta distribution operative at $\boldsymbol{x} = \boldsymbol{x}^S \in \mathcal{D}^S$. With the fundamental source distribution, Eq. (11.34) boils down to

$$\hat{V}_1^{\mathrm{R}}(s)\hat{I}_1^{\mathrm{T}}(s) - \hat{V}_1^{\mathrm{T}}(s)\hat{I}_1^{\mathrm{R}}(s) - \hat{V}_2^{\mathrm{R}}(s)\hat{I}_2^{\mathrm{T}}(s) + \hat{V}_2^{\mathrm{T}}(s)\hat{I}_2^{\mathrm{R}}(s)$$

$$\simeq -\hat{\imath}^{\mathrm{R}}(s) \, \boldsymbol{\ell}^{\mathrm{R}} \cdot \hat{\boldsymbol{E}}^{\mathrm{T}}(\boldsymbol{x}^S, s) \qquad (11.36)$$

thereby replacing the integrals of the excitation field along the line (see Eq. (11.17)) by the radiated electric-field strength observed at $\boldsymbol{x} = \boldsymbol{x}^S$ in the testing state (T). The interaction quantity applying to a magnetic-dipole source representing a small-loop antenna can be found in the same way.

CHAPTER 12

EM PLANE-WAVE INDUCED THÉVENIN'S VOLTAGE ON TRANSMISSION LINES

The main purpose of this chapter is to provide illustrative applications of the EM-reciprocity–based coupling models introduced in chapter 11. For the sake of simplicity, we shall start with the description of Thévenin's voltages induced by a uniform EM plane wave impinging on an idealized model of a transmission line above the perfect ground. Upon analyzing this specific example, it will be shown that the coupling models represented by Eqs. (11.17) and (11.32) are, indeed, fully equivalent. Furthermore, in order to demonstrate that the reciprocity-based coupling models are readily applicable to more complex problem configurations, we will subsequently closely analyze the EM plane-wave induced Thévenin-voltage response of a thin PEC trace on a grounded slab. The chapter is finally concluded by illustrative numerical examples.

12.1 TRANSMISSION LINE ABOVE THE PERFECT GROUND

We shall next analyze the voltage response at the terminals of a transmission that is induced by a uniform EM plane wave (see Figure 12.1). For the sake of simplicity, we assume that the transmission line is uniform and is located along $\{x_1 < x < x_2, y = y_0, z = z_0 > 0\}$ above a PEC ground plane in the homogeneous, isotropic, and loss-free half-space $\mathcal{D}_+^\infty$. EM properties of the latter are described by electric permittivity ϵ_0 and magnetic permeability μ_0 with the corresponding EM wave speed $c_0 = (\epsilon_0\mu_0)^{-1/2} > 0$. The length of the transmission line is denoted by $L = x_2 - x_1 > 0$. In the ensuing sections, we shall provide explicit expressions for Thévenin's voltages induced across

Time-Domain Electromagnetic Reciprocity in Antenna Modeling, First Edition. Martin Štumpf.
© 2020 by The Institute of Electrical and Electronics Engineers, Inc. Published 2020 by John Wiley & Sons, Inc.

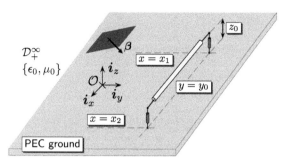

FIGURE 12.1. A plane-wave excited transmission plane over a PEC ground.

the transmission-line ports. With the equivalent Thévenin-voltage generators at our disposal, the induced voltage response for arbitrary load conditions readily follows.

12.1.1   Thévenin's Voltage at $x = x_1$

To arrive at a closed-from expression for the Thévenin equivalent voltage genera-tor at $x = x_1$, the transmission-line Eqs. (11.24) and (11.25) with x-independent (per-unit-length) parameters (see Eqs. (11.6) and (11.7)) are solved subject to the impulsive excitation at $x = x_1$ and the matched load at $x = x_2$, that is

$$\hat{I}_1^{\mathrm{T}}(s) = 1 \tag{12.1}$$

$$\hat{V}_2^{\mathrm{T}}(s) = Y^{\mathrm{c}} \hat{I}_2^{\mathrm{T}}(s) \tag{12.2}$$

where the characteristic admittance Y^{c} has been defined in Eq. (11.9). The testing electric-current distribution then follows as

$$I^{\mathrm{T}}(x, s) = \exp[-s(x - x_1)/c_0] \tag{12.3}$$

for all $\{x_1 \leq x \leq x_2\}$, thus describing a wave propagating in the positive x-direction away from its unit source at $x = x_1$. If, in addition, the analyzed transmission line is in the (actual) receiving situation characteristically terminated at $x = x_2$, that is $\hat{V}_2^{\mathrm{R}}(s) = Y^{\mathrm{c}} \hat{I}_2^{\mathrm{R}}(s)$, then the starting reciprocity relation (11.17) has the following form

$$\hat{V}_1^{\mathrm{G}}(s) \simeq -\int_{x=x_1}^{x_2} \hat{E}_x^{\mathrm{e}}(x, y_0, z_0, s) \exp[-s(x - x_1)/c_0] \mathrm{d}x$$
$$-\int_{z=0}^{z_0} \hat{E}_z^{\mathrm{e}}(x_1, y_0, z, s) \mathrm{d}z + \exp(-sL/c_0) \int_{z=0}^{z_0} \hat{E}_z^{\mathrm{e}}(x_2, y_0, z, s) \mathrm{d}z \tag{12.4}$$

where $\hat{V}_1^G(s)$ is the Thévenin-voltage strength that we look for. To evaluate the voltage response from Eq. (12.4), it is further necessary to specify the excitation-field distribution along the line. To that end, we may use Eq. (11.29) and get

$$\hat{E}_x^e(x, y_0, z_0, s) = -\alpha_x \hat{e}^i(s) \exp[-s(\beta_x x + \beta_y y_0)/c_0]$$
$$\times [\exp(s\beta_z z_0/c_0) - \exp(-s\beta_z z_0/c_0)] \qquad (12.5)$$

$$\hat{E}_z^e(x_{1,2}, y_0, z, s) = \alpha_z \hat{e}^i(s) \exp[-s(\beta_x x_{1,2} + \beta_y y_0)/c_0]$$
$$\times [\exp(s\beta_z z/c_0) + \exp(-s\beta_z z/c_0)] \qquad (12.6)$$

Substituting Eq. (12.5) in the first integral on the right-hand side of (12.4), we find

$$\int_{x=x_1}^{x_2} \hat{E}_x^e(x, y_0, z_0, s) \exp[-s(x - x_1)/c_0] dx$$
$$= -\frac{c_0 \alpha_x \hat{e}^i(s)}{s} \frac{1 - \exp[-sL(1 + \beta_x)/c_0]}{1 + \beta_x} \exp[-s(\beta_x x_1 + \beta_y y_0)/c_0]$$
$$\times [\exp(s\beta_z z_0/c_0) - \exp(-s\beta_z z_0/c_0)] \qquad (12.7)$$

where, in line with the concept of characteristic impedance (see Eq. (11.3)), we further take $sz_0/c_0 \downarrow 0$, which leads to

$$\int_{x=x_1}^{x_2} \hat{E}_x^e(x, y_0, z_0, s) \exp[-s(x - x_1)/c_0] dx$$
$$= -2z_0 \hat{e}^i(s) \alpha_x \beta_z \frac{1 - \exp[-sL(1 + \beta_x)/c_0]}{1 + \beta_x}$$
$$\times \exp[-s(\beta_x x_1 + \beta_y y_0)/c_0] + O(s^2 z_0^2/c_0^2) \qquad (12.8)$$

Similarly, we next proceed with the remaining terms describing the contributions from the transmission-line terminations along $\{x = x_{1,2}, y = y_0, 0 < z < z_0\}$. Hence, the integration of the vertical component of excitation electric field (see Eq. (12.7)) yields

$$\int_{z=0}^{z_0} \hat{E}_z^e(x_{1,2}, y_0, z, s) dz = [c_0 \alpha_z \hat{e}^i(s)/s] \exp[-s(\beta_x x_{1,2} + \beta_y y_0)/c_0]$$
$$\times [\exp(s\beta_z z_0/c_0) - \exp(-s\beta_z z_0/c_0)]/\beta_z \qquad (12.9)$$

whose dominant term as $sz_0/c_0 \downarrow 0$ easily follows, that is

$$\int_{z=0}^{z_0} \hat{E}_z^e(x_{1,2}, y_0, z, s) dz = 2z_0 \hat{e}^i(s) \alpha_z \exp[-s(\beta_x x_{1,2} + \beta_y y_0)/c_0]$$
$$+ O(s^2 z_0^2/c_0^2) \qquad (12.10)$$

Consequently, making use of Eqs. (12.8) and (12.10) in Eq. (12.4), we end up with

$$\hat{V}_1^G(s) \simeq -2z_0\hat{e}^i(s)[\alpha_z(1+\beta_x)-\alpha_x\beta_z]$$
$$\times \frac{1-\exp[-sL(1+\beta_x)/c_0]}{1+\beta_x}\exp[-s(\beta_x x_1+\beta_y y_0)/c_0] \quad (12.11)$$

as $sz_0/c_0 \downarrow 0$. The latter expression can be further simplified for short transmission lines using $1-\exp[-sL(1+\beta_x)/c_0] = sL(1+\beta_x)/c_0 + O(s^2L^2/c_0^2)$ as $sL/c_0 \downarrow 0$, that is

$$\hat{V}_1^G(s) \simeq -2z_0Ls\hat{e}^i(s)c_0^{-1}[\alpha_z(1+\beta_x)-\alpha_x\beta_z]$$
$$\times \exp[-s(\beta_x x_1+\beta_y y_0)/c_0] + O(s^2L^2/c_0^2) \quad (12.12)$$

Finally, the TD counterparts of Eqs. (12.11) and (12.12) immediately follow as

$$V_1^G(t) \simeq -2z_0[\alpha_z-\alpha_x\beta_z/(1+\beta_x)]$$
$$\times \{e^i[t-(\beta_x x_1+\beta_y y_0)/c_0] - e^i[t-(\beta_x x_2+\beta_y y_0)/c_0 - L/c_0]\} \quad (12.13)$$

as $z_0/c_0 t_w \downarrow 0$ and

$$V_1^G(t) \simeq -2z_0(L/c_0)[\alpha_z(1+\beta_x)-\alpha_x\beta_z]$$
$$\times \partial_t e^i[t-(\beta_x x_1+\beta_y y_0)/c_0] \quad (12.14)$$

as $L/c_0 t_w \downarrow 0$, respectively, where $c_0 t_w$ denotes the spatial support of the plane-wave signature. The result given by Eq. (12.11) could also be deduced from the literature on the subject (see [35], for instance).

It is further demonstrated that the induced voltage response could also be found from the alternative formulation (11.32) applying to the plane-wave incidence. To that end, we shall evaluate the far-field radiated amplitude with the aid of Eq. (11.30). Rewriting the latter for the image-source equivalent (see Figure 11.15) under the thin-wire approximation, we find

$$\boldsymbol{\alpha}\cdot\hat{\boldsymbol{E}}^{T;\infty}(-\boldsymbol{\beta},s) = \alpha_x\hat{E}_x^{T;\infty}(-\boldsymbol{\beta},s)|_{z=\pm z_0} + \alpha_z\hat{E}_z^{T;\infty}(-\boldsymbol{\beta},s)|_{x=x_{1,2}}$$
$$= -s\mu_0\,\boldsymbol{\alpha}\cdot\int_{\boldsymbol{x}\in\mathcal{L}\cup\mathcal{L}^*}\exp(-s\boldsymbol{\beta}\cdot\boldsymbol{x}/c_0)\hat{I}^T(\boldsymbol{x},s)\mathrm{d}x \quad (12.15)$$

where $\mathcal{L}\cup\mathcal{L}^*$ refers to the integration along the line and its image. Moreover, the testing-current distribution corresponding to the voltage response at $x=x_1$ has been given in Eq. (12.3). Hence, taking into account the (opposite) orientation of the horizontal electric currents along the line and its image, we get

$$\alpha_x \hat{E}_x^{T;\infty}(-\beta, s)|_{z=\pm z_0} = -s\mu_0 \alpha_x [\exp(-s\beta_z z_0/c_0) - \exp(s\beta_z z_0/c_0)]$$
$$\times \int_{x=x_1}^{x_2} \exp[-s(\beta_x x + \beta_y y_0)/c_0] \exp[-s(x - x_1)/c_0] dx$$
(12.16)

Similarly, for the vertical sections, we find

$$\alpha_z \hat{E}_z^{T;\infty}(-\beta, s)|_{x=x_{1,2}} = \mp s\mu_0 \alpha_z \hat{I}_{1,2}^{T}(s) \exp[-s(\beta_x x_{1,2} + \beta_y y_0)/c_0]$$
$$\times \int_{z=0}^{z_0} [\exp(-s\beta_z z/c_0) + \exp(s\beta_z z/c_0)] dz \quad (12.17)$$

Upon inspection of Eq. (12.16) with (12.5), we next get

$$\hat{e}^{i}(s)\alpha_x \hat{E}_x^{T;\infty}(-\beta, s)|_{z=\pm z_0}$$
$$= -s\mu_0 \int_{x=x_1}^{x_2} \hat{E}_x^{e}(x, y_0, z_0, s) \exp[-s(x - x_1)/c_0] dx$$
(12.18)

while comparing Eq. (12.17) with (12.6) yields

$$\hat{e}^{i}(s)\alpha_z \hat{E}_z^{T;\infty}(-\beta, s)|_{x=x_{1,2}}$$
$$= \mp s\mu_0 \hat{i}_{1,2}^{T}(s) \int_{z=0}^{z_0} \hat{E}_z^{e}(x_{1,2}, y_0, z, s) dz$$
(12.19)

The latter equations provide the link between the radiated far-field amplitudes and the excitation field along the line. Accordingly, making use of Eqs. (12.18) and (12.19) in Eq. (12.4), we end up with

$$\hat{V}_1^{G}(s) \simeq \hat{e}^{i}(s)[\alpha_x \hat{E}_x^{T;\infty}(-\beta, s) + \alpha_z \hat{E}_z^{T;\infty}(-\beta, s)|_{x=x_1}$$
$$+ \alpha_z \hat{E}_z^{T;\infty}(-\beta, s)|_{x=x_2}]/s\mu_0$$
(12.20)

thus proving the validity of the reciprocity-based coupling model (11.32) for the Thévenin-voltage response at $x = x_1$.

12.1.2 Thévenin's Voltage at $x = x_2$

For the sake of completeness, we shall next derive closed-form expressions for the induced Thévenin's voltage at $x = x_2$. For that purpose, we choose the testing electric-current distribution that is excited by the impulsive pulse at $x = x_2$ on the matched transmission line, that is (see Eqs. (12.1) and (12.2))

$$\hat{I}_2^{T}(s) = 1$$
(12.21)

$$\hat{V}_1^{\mathrm{T}}(s) = -Y^{\mathrm{c}}\hat{I}_1^{\mathrm{T}}(s) \tag{12.22}$$

Solving the corresponding transmission-line equations under the conditions (12.21) and (12.22), we get a wave propagating in the negative x-direction, that is

$$I^{\mathrm{T}}(x,s) = \exp[-s(x_2 - x)/c_0] \tag{12.23}$$

for all $\{x_1 \le x \le x_2\}$. Assuming the characteristic termination at $x = x_1$ also in the receiving situation, Eq. (11.17) boils down to

$$\hat{V}_2^{\mathrm{G}}(s) \simeq \int_{x=x_1}^{x_2} \hat{E}_x^{\mathrm{e}}(x, y_0, z_0, s) \exp[-s(x_2 - x)/c_0]\mathrm{d}x$$

$$+ \exp(-sL/c_0)\int_{z=0}^{z_0} \hat{E}_z^{\mathrm{e}}(x_1, y_0, z, s)\mathrm{d}z - \int_{z=0}^{z_0} \hat{E}_z^{\mathrm{e}}(x_2, y_0, z, s)\mathrm{d}z \tag{12.24}$$

where $\hat{V}_2^{\mathrm{G}}(s)$ denotes the equivalent Thévenin voltage at $x = x_2$. Making use of Eqs. (12.5) and (12.6) in Eq. (12.24) and carrying out the integrals analytically, we end up with

$$\hat{V}_2^{\mathrm{G}}(s) \simeq 2z_0\hat{e}^{\mathrm{i}}(s)[-\alpha_z(1-\beta_x) - \alpha_x\beta_z]$$

$$\times \frac{1-\exp[-sL(1-\beta_x)/c_0]}{1-\beta_x}\exp[-s(\beta_x x_2 + \beta_y y_0)/c_0] \tag{12.25}$$

as $sz_0/c_0 \downarrow 0$. Again, the dominant term for short lines easily follows as

$$\hat{V}_2^{\mathrm{G}}(s) \simeq 2z_0 Ls\hat{e}^{\mathrm{i}}(s)c_0^{-1}[-\alpha_z(1-\beta_x) - \alpha_x\beta_z]$$

$$\times \exp[-s(\beta_x x_2 + \beta_y y_0)/c_0] + O(s^2 L^2/c_0^2) \tag{12.26}$$

as $sL/c_0 \downarrow 0$. The latter results can be transformed to the TD, and we finally get (cf. Eqs. (12.13) and (12.14))

$$V_2^{\mathrm{G}}(t) \simeq 2z_0[-\alpha_z - \alpha_x\beta_z/(1-\beta_x)]$$

$$\times \{e^{\mathrm{i}}[t - (\beta_x x_2 + \beta_y y_0)/c_0] - e^{\mathrm{i}}[t - (\beta_x x_1 + \beta_y y_0)/c_0 - L/c_0]\} \tag{12.27}$$

as $z_0/c_0 t_{\mathrm{w}} \downarrow 0$ and

$$V_2^{\mathrm{G}}(t) \simeq 2z_0(L/c_0)[-\alpha_z(1-\beta_x) - \alpha_x\beta_z]$$

$$\times \partial_t e^{\mathrm{i}}[t - (\beta_x x_2 + \beta_y y_0)/c_0] \tag{12.28}$$

as $L/c_0 t_{\mathrm{w}} \downarrow 0$, which completes the analysis of the plane-wave incidence on the transmission line. Using the strategy pursued in the previous section, it can be demonstrated that Eq. (12.25) can be derived from the corresponding radiation characteristics. This task is, again, left as an exercise for the reader.

12.2 NARROW TRACE ON A GROUNDED SLAB

In this section, it is demonstrated that the reciprocity-based coupling model is readily capable of providing the EM plane-wave excited response of a narrow straight printed circuit board (PCB) trace (see Figure 12.2). The PEC trace under consideration is located just on the grounded slab along $\{x_1 < x < x_2, y = y_0, z = 0\}$. The PEC ground plane extends over $\{-\infty < x < \infty, -\infty < y < \infty, z = -d\}$, where $d > 0$ denotes the thickness of the layer. The slab shows a contrast in its EM properties with respect to the homogeneous, isotropic, and loss-free embedding in its (real-valued, and positive) electric permittivity ϵ_1 only. The corresponding EM wave speed is denoted by $c_1 = (\epsilon_1 \mu_1)^{-1/2} > 0$.

As the first step of the analysis that follows, we shall rewrite the starting reciprocity relation (11.17) to the form complying with the problem configuration at hand, that is

$$\hat{V}_1^R(s)\hat{I}_1^T(s) - \hat{V}_1^T(s)\hat{I}_1^R(s) - \hat{V}_2^R(s)\hat{I}_2^T(s) + \hat{V}_2^T(s)\hat{I}_2^R(s)$$

$$\simeq -\int_{x=x_1}^{x_2} \hat{E}_x^e(x, y_0, 0, s)\hat{I}^T(x, s)\mathrm{d}x$$

$$-\hat{I}_1^T(s)\int_{z=-d}^{0} \hat{E}_z^e(x_1, y_0, z, s)\mathrm{d}z + \hat{I}_2^T(s)\int_{z=-d}^{0} \hat{E}_z^e(x_2, y_0, z, s)\mathrm{d}z \quad (12.29)$$

Apparently, the right-hand side of Eq. (12.29) calls for the horizontal component of the excitation field along the trace as well as for its vertical component inside the slab.

As in the previous section, the PCB structure is excited by a uniform EM plane wave as defined by Eq. (11.28). As the EM reflection and transmission properties of a dielectric half-space depend on the polarization of the incident wave, the excitation field as specified by Eq. (11.29) is no longer applicable. Instead, the excitation field is decomposed into its E-polarized (horizontal) and H-polarized (vertical) components denoted by superscripts ^{E} and ^{H}, respectively, that is

$$\hat{E}^e(x, s) = \hat{E}^{e;E}(x, s)\sin(\alpha) + \hat{E}^{e;H}(x, s)\cos(\alpha) \quad (12.30)$$

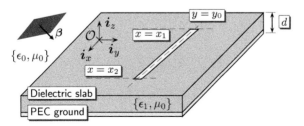

FIGURE 12.2. A plane-wave excited trace on a grounded dielectric slab.

for $\{0 \leq \alpha \leq \pi/2\}$. Consequently, the horizontal component of the excitation field has the following form

$$\hat{E}_\parallel^e(\boldsymbol{x}, s) = [\alpha_\parallel^E \sin(\alpha) + \alpha_\parallel^H \cos(\alpha)]\hat{e}^i(s) \exp[-s(\boldsymbol{\beta}_\parallel + \boldsymbol{\beta}_\perp) \cdot \boldsymbol{x}/c_0]$$
$$+ [\hat{R}^E \alpha_\parallel^E \sin(\alpha) - \hat{R}^H \alpha_\parallel^H \cos(\alpha)]\hat{e}^i(s) \exp[-s(\boldsymbol{\beta}_\parallel - \boldsymbol{\beta}_\perp) \cdot \boldsymbol{x}/c_0] \tag{12.31}$$

in $z > 0$, where the (generalized) reflection coefficients read

$$\hat{R}^E = \frac{r^E - \exp[-(s/c_1)2d\cos(\theta^t)]}{1 - r^E \exp[-(s/c_1)2d\cos(\theta^t)]} \tag{12.32}$$

$$\hat{R}^H = \frac{r^H + \exp[-(s/c_1)2d\cos(\theta^t)]}{1 + r^H \exp[-(s/c_1)2d\cos(\theta^t)]} \tag{12.33}$$

in which

$$r^E = \frac{\eta_0 \cos(\theta) - \eta_1 \cos(\theta^t)}{\eta_0 \cos(\theta) + \eta_1 \cos(\theta^t)} \tag{12.34}$$

$$r^H = \frac{\zeta_0 \cos(\theta) - \zeta_1 \cos(\theta^t)}{\zeta_0 \cos(\theta) + \zeta_1 \cos(\theta^t)} \tag{12.35}$$

with $\eta_{0,1} = (\epsilon_{0,1}/\mu_{0,1})^{1/2}$, $\zeta_{0,1} = \eta_{0,1}^{-1} = (\mu_{0,1}/\epsilon_{0,1})^{1/2}$ and $\sin(\theta)/c_0 = \sin(\theta^t)/c_1$ according to Snell's law. A similar expression applies to the vertical component that has no E-polarized part, that is

$$\hat{E}_\perp^e(\boldsymbol{x}, s) = \alpha_\perp^H \cos(\alpha)\hat{e}^i(s) \exp[-s(\boldsymbol{\beta}_\parallel + \boldsymbol{\beta}_\perp) \cdot \boldsymbol{x}/c_0]$$
$$+ \hat{R}^H \alpha_\perp^H \cos(\alpha)\hat{e}^i(s) \exp[-s(\boldsymbol{\beta}_\parallel - \boldsymbol{\beta}_\perp) \cdot \boldsymbol{x}/c_0] \tag{12.36}$$

for $z > 0$. Finally, the required excitation field inside the layer can be written as

$$\hat{E}_\perp^e(\boldsymbol{x}, s) = \alpha_\perp^H \cos(\alpha)\hat{e}^i(s)(\epsilon_0/\epsilon_1)\hat{T}^H \exp(-s\boldsymbol{\beta}_\parallel \cdot \boldsymbol{x}/c_0)$$
$$\times \{\exp[-sz\cos(\theta^t)/c_1] + \exp[-s(z + 2d)\cos(\theta^t)/c_1]\} \tag{12.37}$$

for $\{-d < z < 0\}$, where the transmission coefficient follows as

$$\hat{T}^H = \frac{1 + r^H}{1 + r^H \exp[-(s/c_1)2d\cos(\theta^t)]} \tag{12.38}$$

It is straightforward to find the limit $z \downarrow 0$ of Eq. (12.36) and get

$$\lim_{z \downarrow 0} \hat{E}_\perp^e(\boldsymbol{x}, s) = \alpha_\perp^H \cos(\alpha)\hat{e}^i(s)(1 + \hat{R}^H) \exp(-s\boldsymbol{\beta}_\parallel \cdot \boldsymbol{x}/c_0) \tag{12.39}$$

while from Eq. (12.37), we may find the limit from below of the interface, that is

$$\lim_{z\uparrow 0} \hat{\boldsymbol{E}}^{\mathrm{e}}_{\perp}(\boldsymbol{x}, s) = (\epsilon_0/\epsilon_1)\boldsymbol{\alpha}^{\mathrm{H}}_{\perp}\cos(\alpha)$$

$$\times \hat{e}^{\mathrm{i}}(s)(1 + \hat{R}^{\mathrm{H}})\exp(-s\boldsymbol{\beta}_{\parallel}\cdot\boldsymbol{x}/c_0) \qquad (12.40)$$

where we used $1 + \hat{R}^{\mathrm{H}} = \hat{T}^{\mathrm{H}}\{1 + \exp[-s2d\cos(\theta^t)/c_1]\}$. The limiting values (12.39) and (12.40) clearly show that the normal component of the excitation electric-field strength exhibits, when crossing the interface, the jump in magnitude that is proportional to the electric contrast ratio ϵ_0/ϵ_1.

The excitation field describing the EM plane-wave response of a grounded slab will be next expressed in spherical coordinates via

$$\boldsymbol{\beta} = \cos(\phi)\sin(\theta)\boldsymbol{i}_x + \sin(\phi)\sin(\theta)\boldsymbol{i}_y - \cos(\theta)\boldsymbol{i}_z \qquad (12.41)$$

$$\boldsymbol{\alpha}^{\mathrm{E}} = \sin(\phi)\boldsymbol{i}_x - \cos(\phi)\boldsymbol{i}_y \qquad (12.42)$$

$$\boldsymbol{\alpha}^{\mathrm{H}} = \cos(\phi)\cos(\theta)\boldsymbol{i}_x + \sin(\phi)\cos(\theta)\boldsymbol{i}_y + \sin(\theta)\boldsymbol{i}_z \qquad (12.43)$$

where ϕ and θ denote the azimuthal and polar angles, respectively. Apparently, we have $\boldsymbol{\alpha}^{\mathrm{E,H}}\cdot\boldsymbol{\beta} = 0$.

12.2.1 Thévenin's Voltage at $x = x_1$

Again, we begin with the description of Thévenin's voltage generator at $x = x_1$. To that end, Eq. (12.4) is modified such that it applies to the PCB configuration shown in Figure 12.2, that is

$$\hat{V}^{\mathrm{G}}_1(s) \simeq -\int_{x=x_1}^{x_2}\hat{E}^{\mathrm{e}}_x(x, y_0, 0, s)\exp[-s(x - x_1)/c]\mathrm{d}x$$

$$-\int_{z=-d}^{0}\hat{E}^{\mathrm{e}}_z(x_1, y_0, z, s)\mathrm{d}z$$

$$+ \exp(-sL/c)\int_{z=-d}^{0}\hat{E}^{\mathrm{e}}_z(x_2, y_0, z, s)\mathrm{d}z \qquad (12.44)$$

where c is the EM wave speed of pulse propagating along the PCB trace. First, we evaluate the integral along the (horizontal) transmission line. The corresponding excitation EM field immediately follows from Eq. (12.31) as

$$\hat{E}^{\mathrm{e}}_x(x, y_0, 0, s) = \hat{e}^{\mathrm{i}}(s)[\alpha^{\mathrm{E}}_x(1 + \hat{R}^{\mathrm{E}})\sin(\alpha) + \alpha^{\mathrm{H}}_x(1 - \hat{R}^{\mathrm{H}})\cos(\alpha)]$$

$$\times \exp[-s(\beta_x x + \beta_y y_0)/c_0] \qquad (12.45)$$

In accordance with Eq. (12.44), the horizontal excitation-field component is integrated along the trace, which leads to (cf. Eq. (12.7))

$$
\int_{x=x_1}^{x_2} \hat{E}_x^e(x, y_0, 0, s) \exp[-s(x-x_1)/c] dx
$$

$$
= \frac{c_0 \hat{e}^i(s)}{s} [\alpha_x^E (1 + \hat{R}^E) \sin(\alpha) + \alpha_x^H (1 - \hat{R}^H) \cos(\alpha)]
$$

$$
\times \frac{1 - \exp[-sL(\mathcal{N} + \beta_x)/c_0]}{\mathcal{N} + \beta_x} \exp[-s(\beta_x x_1 + \beta_y y_0)/c_0] \qquad (12.46)
$$

where we have defined $\mathcal{N} \triangleq c_0/c \geq 1$. Under the assumption that the slab is relatively thin, we may simplify Eq. (12.46) via the following expansions

$$
1 + \hat{R}^E = (2sd/c_0) \cos(\theta) + O(s^2 d^2/c_1^2) \qquad (12.47)
$$

$$
1 - \hat{R}^H = (2sd/c_0)[N^2 - \sin^2(\theta)]/N^2 \cos(\theta) + O(s^2 d^2/c_1^2) \qquad (12.48)
$$

as $sd/c_1 \downarrow 0$, where $N \triangleq c_0/c_1 \geq 1$ (do not confuse with $\mathcal{N} = c_0/c$). Considering the dominant terms and using Eqs. (12.42) and (12.43), we finally arrive at

$$
\int_{x=x_1}^{x_2} \hat{E}_x^e(x, y_0, 0, s) \exp[-s(x-x_1)/c] dx
$$

$$
= 2d\hat{e}^i(s) \left[\sin(\phi) \cos(\theta) \sin(\alpha) + \frac{N^2 - \sin^2(\theta)}{N^2} \cos(\phi) \cos(\alpha) \right]
$$

$$
\times \frac{1 - \exp[-sL(\mathcal{N} + \beta_x)/c_0]}{\mathcal{N} + \beta_x} \exp[-s(\beta_x x_1 + \beta_y y_0)/c_0]
$$

$$
+ O(s^2 d^2/c_1^2) \qquad (12.49)
$$

as $sd/c_1 \downarrow 0$. In the step that follows, we shall integrate the normal component of the excitation field as given in Eq. (12.37) in the dielectric layer. The integration can be, again, carried out in closed form. This way leads to (cf. Eq. (12.9))

$$
\int_{z=-d}^{0} \hat{E}_z^e(x_{1,2}, y_0, z, s) dz = [c_1 \alpha_z^H \cos(\alpha) \hat{e}^i(s)/s]
$$

$$
\times (\epsilon_0/\epsilon_1) \hat{T}^H \exp[-s(\beta_x x_{1,2} + \beta_y y_0)/c_0]
$$

$$
\times \{1 - \exp[-2sd\cos(\theta^t)/c_1]\}/\cos(\theta^t) \qquad (12.50)
$$

from which we extract the dominant term as $sd/c_1 \downarrow 0$ and find

$$
\int_{z=-d}^{0} \hat{E}_z^e(x_{1,2}, y_0, z, s) dz = 2d\hat{e}^i(s)\alpha_z^H \cos(\alpha)(\epsilon_0/\epsilon_1)
$$

$$
\times \exp[-s(\beta_x x_{1,2} + \beta_y y_0)/c_0] + O(sd/c_1) \qquad (12.51)
$$

as $sd/c_1 \downarrow 0$. Consequently, employing Eqs. (12.49) and (12.51) in the expression for the Thévenin voltage (12.44), we end up with (cf. Eq. (12.11))

$$
\hat{V}_1^G(s) \simeq -2d\hat{e}^i(s)
$$
$$
\times [\sin(\phi)\cos(\theta)\sin(\alpha) + \cos(\phi)\cos(\alpha) + (\mathcal{N}/N^2)\sin(\theta)\cos(\alpha)]
$$
$$
\times \frac{1 - \exp[-sL(\mathcal{N} + \beta_x)/c_0]}{\mathcal{N} + \beta_x} \exp[-s(\beta_x x_1 + \beta_y y_0)/c_0] \qquad (12.52)
$$

where we used $N^2 = \epsilon_1/\epsilon_0$. Expression (12.52) attains a simpler form for relatively short traces using the Taylor expansion around $sL/c = 0$. Thus, we find (cf. Eq. (12.12))

$$
\hat{V}_1^G(s) \simeq -2dLs\hat{e}^i(s)c_0^{-1}
$$
$$
\times [\sin(\phi)\cos(\theta)\sin(\alpha) + \cos(\phi)\cos(\alpha) + (\mathcal{N}/N^2)\sin(\theta)\cos(\alpha)]
$$
$$
\times \exp[-s(\beta_x x_1 + \beta_y y_0)/c_0] + O(s^2 L^2/c^2) \qquad (12.53)
$$

as $sL/c \downarrow 0$. Finally, transforming expressions (12.52) and (12.53) to TD, we get

$$
V_1^G(t) \simeq -2d
$$
$$
\times \frac{\sin(\phi)\cos(\theta)\sin(\alpha) + \cos(\phi)\cos(\alpha) + (\mathcal{N}/N^2)\sin(\theta)\cos(\alpha)}{\mathcal{N} + \cos(\phi)\sin(\theta)}
$$
$$
\times \{e^i[t - (\beta_x x_1 + \beta_y y_0)/c_0] - e^i[t - (\beta_x x_2 + \beta_y y_0)/c_0 - L/c]\} \qquad (12.54)
$$

as $d/c_1 t_w \downarrow 0$, while

$$
V_1^G(t) \simeq -2d(L/c_0)
$$
$$
\times [\sin(\phi)\cos(\theta)\sin(\alpha) + \cos(\phi)\cos(\alpha) + (\mathcal{N}/N^2)\sin(\theta)\cos(\alpha)]
$$
$$
\times \partial_t e^i[t - (\beta_x x_1 + \beta_y y_0)/c_0] \qquad (12.55)
$$

as $L/ct_w \downarrow 0$ and recall that t_w denotes the pulse time width of the plane-wave signature.

In conclusion, we note that the main results of this section represented by Eqs. (12.54) and (12.55) can be understood as extensions of (12.13) and (12.14) regarding the voltage response of a uniform transmission line above a PEC ground. Indeed, to reveal the link just let $N = \mathcal{N} = 1$, $d \to z_0$ in Eqs. (12.54) and (12.55) and use Eqs. (12.41) and (12.43) to show that

$$
[\alpha_z(1 + \beta_x) - \alpha_x \beta_z]^H = [\sin(\theta) + \cos(\phi)]\cos(\alpha) \qquad (12.56)
$$

$$[\alpha_z(1+\beta_x) - \alpha_x\beta_z]^{\mathrm{E}} = \sin(\phi)\cos(\theta)\sin(\alpha) \tag{12.57}$$

where we used the decomposition introduced in Eq. (12.30).

12.2.2 Thévenin's Voltage at $x = x_2$

For the sake of completeness, we next briefly derive closed-form expressions also for Thévenin's voltage strength at $x = x_2$. For that purpose, we first modify Eq. (12.24) accordingly

$$\hat{V}_2^{\mathrm{G}}(s) \simeq \int_{x=x_1}^{x_2} \hat{E}_x^{\mathrm{e}}(x, y_0, 0, s)\exp[-s(x_2-x)/c]\mathrm{d}x$$

$$+ \exp(-sL/c)\int_{z=-d}^{0} \hat{E}_z^{\mathrm{e}}(x_1, y_0, z, s)\mathrm{d}z$$

$$- \int_{z=-d}^{0} \hat{E}_z^{\mathrm{e}}(x_2, y_0, z, s)\mathrm{d}z \tag{12.58}$$

As demonstrated in the preceding subsection, the integrations involving the excitation fields as given by Eqs. (12.31) and (12.37) can be carried out analytically, which after a few steps of straightforward algebra leads to (cf. Eqs. (12.25) and (12.52))

$$\hat{V}_2^{\mathrm{G}}(s) \simeq 2d\hat{e}^{\mathrm{i}}(s)$$

$$\times [\sin(\phi)\cos(\theta)\sin(\alpha) + \cos(\phi)\cos(\alpha) - (\mathcal{N}/N^2)\sin(\theta)\cos(\alpha)]$$

$$\times \frac{1-\exp[-sL(\mathcal{N}-\beta_x)/c_0]}{\mathcal{N}-\beta_x}\exp[-s(\beta_x x_2 + \beta_y y_0)/c_0] \tag{12.59}$$

as $sd/c_1 \downarrow 0$. Consequently, for short traces, the latter expression can be further simplified, and we get

$$\hat{V}_2^{\mathrm{G}}(s) \simeq 2dLs\hat{e}^{\mathrm{i}}(s)c_0^{-1}$$

$$\times [\sin(\phi)\cos(\theta)\sin(\alpha) + \cos(\phi)\cos(\alpha) - (\mathcal{N}/N^2)\sin(\theta)\cos(\alpha)]$$

$$\times \exp[-s(\beta_x x_2 + \beta_y y_0)/c_0] + O(s^2 L^2/c^2) \tag{12.60}$$

as $sL/c \downarrow 0$. In the final step, Eqs. (12.59) and (12.60) are transformed back to the TD, which yields

$$V_2^{\mathrm{G}}(t) \simeq 2d$$

$$\times \frac{\sin(\phi)\cos(\theta)\sin(\alpha) + \cos(\phi)\cos(\alpha) - (\mathcal{N}/N^2)\sin(\theta)\cos(\alpha)}{\mathcal{N} - \cos(\phi)\sin(\theta)}$$

$$\times \{e^{\mathrm{i}}[t - (\beta_x x_2 + \beta_y y_0)/c_0] - e^{\mathrm{i}}[t - (\beta_x x_1 + \beta_y y_0)/c_0 - L/c]\} \tag{12.61}$$

as $d/c_1 t_w \downarrow 0$, while

$$V_2^G(t) \simeq 2d(L/c_0)$$
$$\times [\sin(\phi)\cos(\theta)\sin(\alpha) + \cos(\phi)\cos(\alpha) - (\mathcal{N}/N^2)\sin(\theta)\cos(\alpha)]$$
$$\times \partial_t e^i[t - (\beta_x x_2 + \beta_y y_0)/c_0] \tag{12.62}$$

as $L/ct_w \downarrow 0$. Equations (12.61) and (12.62) can be, again, interpreted as generalizations of (12.27) and (12.28) applying to a transmission line above the PEC ground. With the characteristic impedance of a narrow trace on a microstrip structure at our disposal (see [41, eq. (3.196)], for example), Eqs. (12.54) and (12.55) with (12.61) and (12.62) can serve for calculating the plane-wave induced TD voltage response of a PCB trace for arbitrary loading conditions. This problem has been originally analyzed in the context of EMC in Ref. [42] with the aid of "Baum-Liu-Tesche equations."

ILLUSTRATIVE NUMERICAL EXAMPLE

- Make use of Eq. (12.54) with Eq. (12.61) to calculate the EM plane-wave induced (open-circuited) Thévenin-voltage responses at the both ends of a PCB trace. Subsequently, validate the approximate expressions with the aid of a three-dimensional EM computational tool.

Solution: We shall calculate the induced Thévenin's voltage responses of a PCB trace located on a grounded dielectric slab. The length of the trace is $L = 100\,\text{mm}$, and its width is $w = 0.20\,\text{mm}$. The height of the dielectric slab is $d = 1.50\,\text{mm}$, and its (relative) electric permittivity is $N^2 = \epsilon_1/\epsilon_0 = 4.0$. The microstrip structure is irradiated by a uniform plane-wave whose propagation and polarization vectors can be determined from Eqs. (12.41)–(12.43) with $\phi = 0$, $\theta = \pi/6$ and $\alpha = \pi/4$. The plane-wave pulse shape is defined by Eq. (9.3), where we take $e_m = 1.0\,\text{V/m}$ and $c_0 t_w = 1.0\,L$. Apparently, the height of the slab is sufficiently small with respect to the spatial support of the exciting plane-wave signature, so that Eqs. (12.54) and (12.61) are applicable.

The required Thévenin-voltage response corresponds to an open-circuit voltage with the matched termination at the other end of the line. Accordingly, to calculate the load impedance, we apply [41, eqs. (3.195) and (3.196)] and write

$$\mathcal{N}^2 = (N^2 + 1)/2 + (N^2 - 1)/[2(1 + 12d/w)^{1/2}] \tag{12.63}$$

recalling that $\mathcal{N}^2 = c_0^2/c^2$ corresponds to the effective dielectric constant, and $N^2 = c_0^2/c_1^2$ is equal to the relative electric permittivity of the slab. Subsequently, the characteristic impedance $Z^c = 1/Y^c$ (see Eq. (11.9)) follows from

$$Z^c = (60/\mathcal{N})\ln(8d/w + w/4d) \tag{12.64}$$

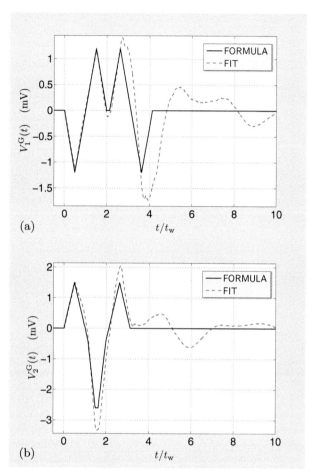

(a)

(b)

FIGURE 12.3. Induced Thévenin's voltage responses as calculated using formulas (12.54) and (12.61) and FIT. The response observed at (a) $x = x_1$ and (b) $x = x_2$.

for $w/d \leq 1$. For the chosen configuration parameters, we approximately get $\mathcal{N}^2 \simeq 2.7$ and $Z^c \simeq 150\ \Omega$. The first parameter is used to find the wave speed of the pulse propagating along the trace, that is $c = c_0/\mathcal{N}$, while the second one is used to match the line in a three-dimensional EM computational tool. For this purpose, we employ the CST Microwave Studio relying on the Finite Integration Technique (FIT). The incident EM plane wave has been chosen such that it hits $\{x = x_{1,2}, y = y_0 = 0, z = 0\}$ at $t = 0$ for $V_{1,2}^G(t)$, respectively. The resulting pulse shapes are shown in Figure 12.3. It is observed that the calculated responses in their early parts correspond to each other very well. Apparent discrepancies appear later, in particular at observation times where the approximate formulas do not predict any field at the terminals. As the (bounded) FIT model consists of the PCB trace placed on a 200 mm × 50 mm × 1.50 mm dielectric box grounded by the (finite) PEC plane, the differences can largely be attributed to the different models and to a non ideal line matching. The differences between the pulses can be significantly reduced by incorporating the (infinite) PEC plane in the FIT model. Overall, the results produced by the approximate closed-form formulas are satisfactory.

CHAPTER 13

VED-INDUCED THÉVENIN'S VOLTAGE ON TRANSMISSION LINES

In chapter 12, it has been shown that expressing the induced voltage response of a transmission line is rather straightforward for a plane-wave incidence. The reason behind the elementary calculations involved is a simple distribution of the excitation field along the line, or, equivalently, a relatively simple form of the far-field radiation characteristics of the line operating as a transmitter. In case that the effect of a disturbing EM field can no longer be approximated by a uniform EM plane, the expression for the excitation field may not be elementary anymore. This is exactly demonstrated in the present chapter, where the VED–induced voltage response of a transmission line is analyzed. For the sake of clarity, the idealized transmission line above the PEC ground is analyzed first. Consequently, a finite ground conductivity is incorporated approximately via the so-called Cooray-Rubinstein formula [26]. The mathematical analysis is carried out with the aid of the CdH method [8].

13.1 TRANSMISSION LINE ABOVE THE PERFECT GROUND

We shall next analyze the voltage response at the terminals of a transmission line that is induced by a VED source located at $(0, 0, h > 0)$ above a planar ground (see Figure 13.1). The corresponding source signature is described by $j(t)$ (in A · m). In this section, the ground is assumed to be EM-impenetrable with $\sigma \to \infty$ (or $\epsilon_1 \to \infty$), thus modeling the PEC ground. We further assume that the transmission line is uniform and is located along $\{x_1 < x < x_2, y = y_0, z = z_0 > 0\}$ above the ground plane in the homogeneous, isotropic, and loss-free half-space $\mathcal{D}_+^\infty$. Again, EM properties of the upper half-space are described by electric permittivity ϵ_0 and magnetic permeability μ_0 with the corresponding EM wave speed $c_0 = (\epsilon_0 \mu_0)^{-1/2} > 0$. The length of the transmission line is denoted by $L = x_2 - x_1 > 0$. Employing the EM reciprocity-based coupling model as in sections 12.1 and 12.2, we next derive

Time-Domain Electromagnetic Reciprocity in Antenna Modeling, First Edition. Martin Štumpf.

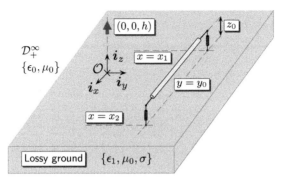

FIGURE 13.1. A VED-excited transmission plane over a lossy ground.

closed-form space-time expressions concerning Thévenin's voltages induced across the ports of the transmission line. To that end, we begin with the analysis of the pertaining excitation EM wave fields (see Eq. (11.12)).

13.1.1 Excitation EM Fields

Closed-form formulas describing the EM field radiated from a VED above the perfect ground are well-known (see [43, eqs. (1)–(3)], for example). As a direct integration of such expressions along the transmission line is not elementary, we next proceed in a different way. Hence, the EM fields radiated from the VED are first represented via [3, eqs. (26.3-1) and (26.3-2)]

$$\hat{E}_x^e(\boldsymbol{x}, s) = \hat{\jmath}(s)\partial_x\partial_z\hat{G}(\boldsymbol{x}, s)/s\epsilon_0 \tag{13.1}$$

$$\hat{E}_y^e(\boldsymbol{x}, s) = \hat{\jmath}(s)\partial_y\partial_z\hat{G}(\boldsymbol{x}, s)/s\epsilon_0 \tag{13.2}$$

$$\hat{E}_z^e(\boldsymbol{x}, s) = -s\mu_0\hat{\jmath}(s)\hat{G}(\boldsymbol{x}, s) + \hat{\jmath}(s)\partial_z^2\hat{G}(\boldsymbol{x}, s)/s\epsilon_0 \tag{13.3}$$

$$\hat{H}_x^e(\boldsymbol{x}, s) = \hat{\jmath}(s)\partial_y\hat{G}(\boldsymbol{x}, s) \tag{13.4}$$

$$\hat{H}_y^e(\boldsymbol{x}, s) = -\hat{\jmath}(s)\partial_x\hat{G}(\boldsymbol{x}, s) \tag{13.5}$$

where $\hat{G}(\boldsymbol{x}, s)$ is the (bounded) solution of the homogeneous, three-dimensional scalar (modified) Helmholtz equation

$$(\partial_x^2 + \partial_y^2 + \partial_z^2 - s^2/c_0^2)\hat{G}(\boldsymbol{x}, s) = 0 \tag{13.6}$$

that is supplemented with the excitation condition accounting for the VED source

$$\lim_{z\downarrow h}\partial_z\hat{G}(\boldsymbol{x}, s) - \lim_{z\uparrow h}\partial_z\hat{G}(\boldsymbol{x}, s) = -\delta(x, y) \tag{13.7}$$

for all $x \in \mathbb{R}$ and $y \in \mathbb{R}$, where $\delta(x, y)$ denotes the two-dimensional Dirac delta distribution operative at $x = y = 0$. Furthermore, on account of the explicit-type boundary condition on the PEC ground, that is

$$\lim_{z \downarrow 0} \{\hat{E}_x^e, \hat{E}_y^e\}(x, s) = 0 \tag{13.8}$$

for all $x \in \mathbb{R}$ and $y \in \mathbb{R}$, Eqs. (13.1) and (13.2) imply that

$$\lim_{z \downarrow 0} \partial_z \hat{G}(x, s) = 0 \tag{13.9}$$

for all $x \in \mathbb{R}$ and $y \in \mathbb{R}$, thereby incorporating the presence of the PEC ground plane. The Helmholtz equation (13.6) subject to conditions (13.7) and (13.9), and the "boundedness condition" applying at $|x| \to \infty$ will be next solved with the aid of the wave slowness integral representation taken in both x- and y-directions.

Transform-Domain Solution

The solution of the Helmholtz equation is expressed via the wave-slowness representation in the following form

$$\hat{G}(x, s) = (s/2\pi i)^2 \int_{\kappa=-i\infty}^{i\infty} d\kappa$$

$$\times \int_{\sigma=-i\infty}^{i\infty} \exp[-s(\kappa x + \sigma y)] \tilde{G}(\kappa, \sigma, z, s) d\sigma \tag{13.10}$$

for $\{s \in \mathbb{R}; s > 0\}$, which entails that $\hat{\partial}_x = -s\kappa$ and $\hat{\partial}_y = -s\sigma$. Under the representation, the transform-domain Helmholtz equation (13.6) reads

$$[\partial_z^2 - s^2 \Gamma_0^2(\kappa, \sigma)]\tilde{G}(\kappa, \sigma, z, s) = 0 \tag{13.11}$$

where $\Gamma_0(\kappa, \sigma) \triangleq (1/c_0^2 - \kappa^2 - \sigma^2)^{1/2}$ with $\mathrm{Re}(\Gamma_0) \geq 0$ is the wave-slowness parameter in the z-direction, while the boundary conditions (13.7) and (13.9) transform to

$$\lim_{z \downarrow h} \partial_z \tilde{G}(\kappa, \sigma, z, s) - \lim_{z \uparrow h} \partial_z \tilde{G}(\kappa, \sigma, z, s) = -1 \tag{13.12}$$

$$\lim_{z \downarrow 0} \partial_z \tilde{G}(\kappa, \sigma, z, s) = 0 \tag{13.13}$$

respectively. Accordingly, the transform-domain solution of Eqs. (13.11)–(13.13) will consist of wave constituents propagating in the (up-going and down-going) directions normal to the ground plane. The general solution that is bounded as $z \to \infty$ can be written as

$$\tilde{G} = A \exp[-s\Gamma_0(z - h)] + B \exp[-s\Gamma_0(z + h)] \tag{13.14}$$

for $\{z \geq h\}$ and

$$\tilde{G} = A \exp[s\Gamma_0(z - h)] + B \exp[-s\Gamma_0(z + h)] \tag{13.15}$$

for $\{0 \leq z \leq h\}$, where A and B are unknown coefficients to be determined from the (transform-domain) boundary conditions (13.12) and (13.13). Hence, making use of (13.14) and (13.15) in (13.12) and (13.13), we get at once

$$A = B = 1/2s\Gamma_0(\kappa, \sigma) \tag{13.16}$$

thus specifying the solution in the transform domain. In view of the terms on the right-hand side of the reciprocity relation (11.17), we use the transform-domain counterparts of Eqs. (13.1) and (13.3) to write down the relevant expressions for the axial electric-field component and the (integral of the) vertical electric-field strength. For the former, this way leads to

$$\tilde{E}_x^e(\kappa, \sigma, z, s) = -[\hat{j}(s)p/2\epsilon_0]$$
$$\times \{\exp[-s\Gamma_0(h - z)] - \exp[-s\Gamma_0(z + h)]\} \tag{13.17}$$

for $\{0 \leq z < h\}$ and

$$\tilde{E}_x^e(\kappa, \sigma, z, s) = [\hat{j}(s)p/2\epsilon_0]$$
$$\times \{\exp[-s\Gamma_0(z - h)] + \exp[-s\Gamma_0(z + h)]\} \tag{13.18}$$

for $\{z > h\}$, while the latter can be written as

$$\int_{z=0}^{z_0} \tilde{E}_z^e(\kappa, \sigma, z, s)\mathrm{d}z = -[\hat{j}(s)(\kappa^2 + \sigma^2)/2s\epsilon_0\Gamma_0^2(\kappa, \sigma)]$$
$$\times \{\exp[-s\Gamma_0(h - z_0)] - \exp[-s\Gamma_0(z_0 + h)]\} \tag{13.19}$$

for $\{0 < z_0 \leq h\}$ and

$$\int_{z=0}^{z_0} \tilde{E}_z^e(\kappa, \sigma, z, s)\mathrm{d}z = -[\hat{j}(s)(\kappa^2 + \sigma^2)/2s\epsilon_0\Gamma_0^2(\kappa, \sigma)]$$
$$\times \{2 - \exp[-s\Gamma_0(z_0 - h)] - \exp[-s\Gamma_0(z_0 + h)]\} \tag{13.20}$$

for $z_0 \geq h$.

13.1.2 Thévenin's Voltage at $x = x_1$

The Thévenin-voltage response at $x = x_1$ is, again, evaluated with the help of Eq. (12.4). In the first step, we will specify the contribution corresponding to the horizontal section of the line along $\{x_1 < x < x_2, y = y_0, z = z_0\}$ (see Figure 13.1). To that end, the wave-slowness representation of Eqs. (13.17) and (13.18) is substituted in the first term on the right-hand side of Eq. (12.4), and in the resulting integral expressions, we change the order of the integrations with respect to x and κ. This leads to elementary inner integrals along $\{x_1 < x < x_2\}$ that can easily be carried out analytically. In this way, we end up with an expression that can be composed of constituents the generic integral representation of which is closely analyzed in section F.1. In the ensuing step, contributions from the vertical sections along $\{x = x_{1,2}, y = y_0, 0 < z < z_0\}$ are found. Accordingly, with reference to the second and third terms on the right-hand side of Eq. (12.4), the transform-domain expressions (13.19) and (13.20) call for their transformation back to the TD. This can be accomplished, again, via the CdH method along the lines detailed in section F.2. Collecting these results, we finally end up with

$$V_1^G(t) \simeq - \, \mathcal{Q}(x_1|x_2, y_0, h - z_0, t) + \mathcal{Q}(x_1|x_2, y_0, z_0 + h, t)$$
$$+ \, \mathcal{V}(x_1, y_0, t) - \mathcal{V}(x_2, y_0, t - L/c_0) \tag{13.21}$$

for $\{0 < z_0 < h\}$ and

$$V_1^G(t) \simeq \mathcal{Q}(x_1|x_2, y_0, z_0 - h, t) + \mathcal{Q}(x_1|x_2, y_0, z_0 + h, t)$$
$$+ \, \mathcal{V}(x_1, y_0, t) - \mathcal{V}(x_2, y_0, t - L/c_0) \tag{13.22}$$

for $z_0 > h$, where

$$\mathcal{Q}(x_1|x_2, y, z, t) = \zeta_0 \, \partial_t j(t)$$
$$*_t \, [\mathcal{I}(x_2, y, z, t - L/c_0) - \mathcal{I}(x_1, y, z, t)] \tag{13.23}$$

with $R(x, y, z) = (x^2 + y^2 + z^2)^{1/2}$, where $\mathcal{I}(x, y, z, t)$ is specified by Eq. (F.19). Next, we have

$$\mathcal{V}(x, y, t) = \mathcal{U}(x, y, h - z_0, t) - \mathcal{U}(x, y, z_0 + h, t) \tag{13.24}$$

for $\{0 < z_0 < h\}$ and

$$\mathcal{V}(x, y, t) = 2 \, \mathcal{U}(x, y, 0, t) - \mathcal{U}(x, y, z_0 - h, t)$$
$$- \, \mathcal{U}(x, y, z_0 + h, t) \tag{13.25}$$

for $z_0 > h$. In Eqs. (13.24) and (13.25), the space-time function $\mathcal{U}$ has the following form

$$\mathcal{U}(x, y, z, t) = \zeta_0 \, \partial_t j(t) *_t \mathcal{J}(x, y, z, t) \tag{13.26}$$

with $\mathcal{J}(x, y, z, t)$ is given in Eq. (F.35).

13.1.3   Thévenin's Voltage at $x = x_2$

We shall proceed in a similar way as in the previous section. Hence, we start from Eq. (12.24), where we substitute the wave-slowness representations of the relevant excitation EM fields. Consequently, employing the results of appendix F, we arrive at

$$V_2^G(t) \simeq - \mathcal{Q}(-x_2| - x_1, y_0, h - z_0, t) + \mathcal{Q}(-x_2| - x_1, y_0, z_0 + h, t)$$
$$+ \mathcal{V}(x_2, y_0, t) - \mathcal{V}(x_1, y_0, t - L/c_0) \tag{13.27}$$

for $\{0 < z_0 < h\}$ and

$$V_2^G(t) \simeq \mathcal{Q}(-x_2| - x_1, y_0, z_0 - h, t) + \mathcal{Q}(-x_2| - x_1, y_0, z_0 + h, t)$$
$$+ \mathcal{V}(x_2, y_0, t) - \mathcal{V}(x_1, y_0, t - L/c_0) \tag{13.28}$$

for $z_0 > h$. In Eqs. (13.27) and (13.28), we use Eqs. (13.23)–(13.25) as defined in the previous section.

13.2   INFLUENCE OF FINITE GROUND CONDUCTIVITY

In the preceding section, we have derived closed-form formulas facilitating the efficient calculation of the VED-induced Thévenin-voltage responses on a transmission line over a PEC ground. In this respect, a natural question arises what is the impact of a finite ground conductivity on the voltage response. This problem is (approximately) analyzed in the present section with the help of the Cooray-Rubinstein formula that has been developed for calculating lightning-induced voltages on overhead transmission lines [26].

The problem configuration under consideration in the present section is shown in Figure 13.1. The transmission line is located in the homogeneous, isotropic, and loss-free half-space $\mathcal{D}_+^\infty$ over the planar ground occupying the lower, homogeneous, and isotropic half-space in $\{-\infty < x < \infty, -\infty < y < \infty, -\infty < z < 0\}$. EM properties of the ground are described by its electric permittivity ϵ_1, magnetic permeability μ_0, and electric conductivity σ. In our approximate model, the presence of the ground will be accounted for via a (linear, time-invariant and local) surface-impedance boundary condition. The analysis is further limited to "ideal lines" along which the pulse propagates with the EM wave speed of $\mathcal{D}_+^\infty$, that is, its wave speed is not affected by the finite ground conductivity. This assumption is for typical lightning-induced voltage calculations justifiable for transmission lines whose length is shorter than about 2.0 km [44].

13.2.1   Excitation EM Fields

To account for the effect of a finite ground conductivity, the explicit-type boundary conditions (13.8) are replaced with surface-impedance boundary conditions (cf. [26, eq. (5)])

$$\lim_{z\downarrow0}\{\hat{F}^e_{,x}, \hat{E}^e_y\}(x, \boldsymbol{o}) = \hat{Z}(s)\lim_{z\downarrow0}\{-\hat{H}^c_y, \hat{H}^c_x\}(x, s) \qquad (13.29)$$

for all $x \in \mathbb{R}$ and $y \in \mathbb{R}$, where $\hat{Z}(s)$ denotes the surface impedance. The latter is assumed to have the following form

$$\hat{Z}(s) = \zeta_1 s/[s(s + \alpha)]^{1/2} \qquad (13.30)$$

with $\zeta_1 = (\mu_0/\epsilon_1)^{1/2} > 0$ and $\alpha = \sigma/\epsilon_1$. Note that $\hat{Z}(s) = 0$ for the PEC ground. Substitution of Eqs. (13.1), (13.2), (13.4), and (13.5) in the impedance boundary condition (13.29) implies that

$$\lim_{z\downarrow0} \partial_z \hat{G}(x, s) = s\epsilon_0 \hat{Z}(s) \lim_{z\downarrow0} \hat{G}(x, s) \qquad (13.31)$$

for all $x \in \mathbb{R}$ and $y \in \mathbb{R}$, which is clearly a generalization of the (Neumann-type) boundary condition (13.9).

To conclude, the problem is now formulated in terms of the Helmholtz equation (13.6) subject to the excitation condition (13.7), the impedance boundary condition (13.31), and the "boundedness condition" at $|x| \to \infty$. The ensuing analysis will be, again, carried out under the wave slowness representation taken in the directions parallel to the ground plane (see Eq. (13.10)).

Transform-Domain Solution

Solving the transform-domain Helmholtz equation (13.11) subject to the excitation condition (13.12) and the transform-domain counterpart of Eq. (13.31), that is

$$\lim_{z\downarrow0} \partial_z \tilde{G}(\kappa, \sigma, z, s) = s\epsilon_0 \hat{Z}(s) \lim_{z\downarrow0} \tilde{G}(\kappa, \sigma, z, s) \qquad (13.32)$$

leads to the solution in the form of Eqs. (13.14) and (13.15) with

$$A = 1/2s\Gamma_0(\kappa, \sigma) \qquad (13.33)$$

$$B = A \tilde{R} \qquad (13.34)$$

where we have defined the transform-domain reflection coefficient

$$\tilde{R} \triangleq [c_0\Gamma_0(\kappa, \sigma) - \hat{Z}(s)/\zeta_0]/[c_0\Gamma_0(\kappa, \sigma) + \hat{Z}(s)/\zeta_0] \qquad (13.35)$$

with $\zeta_0 = (\mu_0/\epsilon_0)^{1/2} > 0$ being the wave impedance of the upper half-space.

In the following (approximate) analysis, we shall rely on assumptions that apply to calculations of lightning-induced voltages on overhead transmission lines [26]. In particular, it has been observed that while the (actual) vertical component of the excitation electric field is not for typical problem configurations significantly affected by the finite ground conductivity, this is not the case for the horizontal component, the impact on which must be accounted for. Consequently, we can make use of the vertical-field contributions from section 13.1 and define the correction (of the horizontal field component) to the result concerning the PEC ground, that is

$$\Delta \tilde{E}_x^e(\kappa, \sigma, z, s)|_{[h, \hat{Z}]} \triangleq \tilde{E}_x^e(\kappa, \sigma, z, s)|_{[h, \hat{Z}]} - \tilde{E}_x^e(\kappa, \sigma, z, s)|_{[h, 0]} \qquad (13.36)$$

where the first and second symbol in the square brackets refers to the height of the VED source and the surface ground impedance, respectively. Under the decomposition, it is straightforward to show that the following relation holds true

$$\Delta \tilde{E}_x^e(\kappa, \sigma, z, s)|_{[h, \hat{Z}]} = -\hat{Z}(s)\tilde{H}_y^e(\kappa, \sigma, 0, s)|_{[z+h, \hat{Z}]} \qquad (13.37)$$

Apparently, as $z \downarrow 0$, Eq. (13.37) on account of Eq. (13.8) boils down to the (transform-domain) impedance boundary condition (13.29). The idea behind the Cooray-Rubinstein model is to simplify the problem by replacing the tangential component of the magnetic field at the ground level by the one pertaining to the PEC ground. In this way, Eq. (13.37) is approximated by

$$\Delta \tilde{E}_x^e(\kappa, \sigma, z, s)|_{[h, \hat{Z}]} \overset{\text{CR}}{\simeq} -\hat{Z}(s)\tilde{H}_y^e(\kappa, \sigma, 0, s)|_{[z+h, 0]} \qquad (13.38)$$

which is, in fact, the (transform-domain) Cooray-Rubinstein formula. Since the transform-domain solution has been determined (see Eqs. (13.33)–(13.35) with Eqs. (13.14) and (13.15)), we may use the transform-domain counterpart of Eq. (13.5) and rewrite the right-hand side of Eq. (13.38) as follows:

$$\Delta \tilde{E}_x^e(\kappa, \sigma, z, s)|_{[h, \hat{Z}]} \overset{\text{CR}}{\simeq} -\hat{j}(s)\hat{Z}(s)[\kappa/\Gamma_0(\kappa, \sigma)] \exp[-s\Gamma_0(z + h)] \qquad (13.39)$$

The latter can hence be viewed as the Cooray-Rubinstein correction to Eqs. (13.17) and (13.18) that accounts for the finite ground conductivity.

13.2.2   Correction to Thévenin's Voltage at $x = x_1$

The slowness representation of the correction term derived in the previous section is next substituted in (the first term on the right-hand side of) Eq. (12.4) to find the corresponding term incorporating the effect of the finite ground conductivity. Assuming the "ideal line" and following the strategy described in section 13.1.2, we will end up with the voltage correction (or the incremental voltage) with respect to the PEC ground that can be represented via Eq. (F.43), that is

$$\Delta V_1^G(t) \overset{\text{CR}}{\simeq} Z(t) *_t \partial_t j(t)$$
$$*_t [\mathcal{K}(x_1, y_0, z_0 + h, t) - \mathcal{K}(x_2, y_0, z_0 + h, t - L/c_0)] \qquad (13.40)$$

where $Z(t)$ denotes the TD original of the s-domain surface impedance (13.30). The TD impedance follows at once using [25, eq. (29.3.49)], that is

$$Z(t) = \zeta_1 \partial_t[\mathsf{I}_0(\alpha t/2)\mathrm{H}(t)]$$
$$= \zeta_1\{\delta(t) - (\alpha/2)[\mathsf{I}_0(\alpha t/2) - \mathsf{I}_1(\alpha t/2)]\mathrm{H}(t)\} \qquad (13.41)$$

where $\mathsf{I}_{0,1}(t) \triangleq I_{0,1}(t)\exp(-t)$ are the (scaled) modified Bessel functions of the first kind (see [25, figure 9.8]). Clearly, we can make use of Eq. (13.41) and rewrite $Z(t) *_t \partial_t j(t)$ as $\partial_t^{-1} Z(t) *_t \partial_t^2 j(t)$ provided that the source signature is a twice differentiable function, and ∂_t^{-1} denotes the time-integration operator (see [28, eq. (20)]). Finally, the result of Eq. (13.40) is added to Eqs. (13.21) and (13.22) to get the total voltage responses of a transmission line above the lossy ground.

13.2.3   Correction to Thévenin's Voltage at $x = x_2$

Following the approach presented in the previous section, we may find an expression similar to Eq. (13.40) that describes the correction to the Thévenin-voltage response at $x = x_2$ as predicted by the Cooray-Rubinstein model. This strategy leads to

$$\Delta V_2^{\mathrm{G}}(t) \overset{\mathrm{CR}}{\simeq} Z(t) *_t \partial_t j(t)$$
$$*_t [\mathcal{K}(-x_2, y_0, z_0 + h, t) - \mathcal{K}(-x_1, y_0, z_0 + h, t - L/c_0)] \quad (13.42)$$

where $\mathcal{K}(x, y, z, t)$ is, again, given by Eq. (F.43). Finally, upon adding Eq. (13.42) to Eqs. (13.27) and (13.28), we end up with the induced, open-circuit voltage across the port at $x = x_2$ of a transmission line above the lossy ground.

ILLUSTRATIVE NUMERICAL EXAMPLE

• Calculate the voltage responses at the both terminals of a matched overhead line as induced by a typical subsequent return stroke.

Solution: The required lightning-induced voltage responses can be calculated using a straightforward modification of the demo MATLAB® code given in appendix L. In particular, the return-stroke channel is viewed as to be composed of VED-sources, a (space-time) distribution of which is described by the modified transmission-line model [45]. Accordingly, the source signature $j(t)$ is at the actual source height h represented via $i(t - h/v)\exp(-h/\lambda)\Delta h$, where $v = 1.30 \cdot 10^8$ m/s is the wave speed of the return stroke, $\lambda = 2.0$ km denotes the decay constant, and Δh is the (relatively short) length of a VED. As the contributions of VEDs along the lightning channel are finally integrated, Δh has, in fact, the meaning of the spatial step of integration. The pulse shape of the current at the base is described as a sum of two functions of the type [45, eq. (2)]

$$i(t) = \frac{I_0}{\eta} \frac{(t/\tau_1)^n}{1 + (t/\tau_1)^n} \exp(-t/\tau_2)H(t) \tag{13.43}$$

with $\eta = \exp[-(\tau_1/\tau_2)(n\tau_2/\tau_1)^{1/n}]$, where we take $I_0 = 10.7$ kA, $\tau_1 = 0.25$ μs, $\tau_2 = 2.5$ μs, and $n = 2$ for the first waveform and $I_0 = 6.5$ kA, $\tau_1 = 2.1$ μs, $\tau_2 = 230$ μs, and $n = 2$ for the second one. It is noted that for $n = 2$, the pulse shape (13.43) is two times differentiable with the starting values $\lim_{t \downarrow 0} i(t) = 0$, $\lim_{t \downarrow 0} \partial_t i(t) = 0$, and $\lim_{t \downarrow 0} \partial_t^2 i(t) = 2I_0/\eta \, \tau_1^2$.

With the definition of the source distribution at our disposal, we may next calculate the voltage induced on the matched transmission line of length $L = 1.0$ km that is placed along $\{x_1 = -L/2 < x < x_2 = L/2, y_0 = L/20, z_0 = L/100\}$ above the ground plane. The EM parameters of the latter are $\epsilon_1 = 10 \, \epsilon_0$ and $\sigma = 10^{-3}$ S/m. Thanks to the symmetry of the problem configuration, the induced voltages at the ends of the transmission line are equal. Moreover, for the analyzed case of the matched line, the induced voltage is simply equal to half of the Thévenin voltage, that is, to $V_{1,2}^G(t)/2$. In this way, we obtain the results shown in Figure 13.2. To illustrate the effect of the ground finite conductivity, we have further included the corresponding voltage response of the line above the PEC ground. The same problem configuration has been thoroughly analyzed in Ref. [44] using a TD finite-difference technique applied to the Agrawal EM-field-to-line coupling model (see section 11.4).

In Refs. [36, 46], it has been then demonstrated that the numerical results presented in Ref. [44] correlate well with the analytical ones.

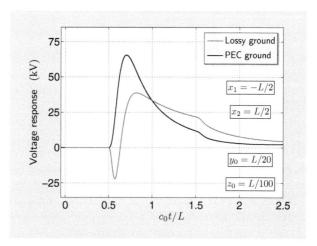

FIGURE 13.2. Induced voltage response across the ports of the matched line. Parameters of lossy ground are $\epsilon_1/\epsilon_0 = 10$ and $\sigma = 10^{-3}$ S/m.

CHAPTER 14

CAGNIARD-DEHOOP METHOD OF MOMENTS FOR PLANAR-STRIP ANTENNAS

In chapter 2, we have shown that the EM reciprocity theorem of the time-convolution type in combination with the classic CdH method can serve for introducing a novel TD integral-equation technique that is capable of analyzing EM scattering and radiation from thin-wire (cylindrical) antennas. In an effort to put forward closed-form and easy-to-implement expressions for the elements of the corresponding TD impedance matrix, we have done away with the (second-order) terms $O(a^2)$ that are negligible as the radius of wire a approaches zero. As the ratio of the radius to the (spatial extent of the) time step influences the stability of the algorithm, it seems instructive, before we proceed with the development of a computational procedure applying to a planar narrow-strip antenna, to take a closer look at the behavior of slowness-domain interactions as the radius of a wire and the width of a strip are approaching zero. As will become clear, such an asymptotic analysis yields a clue to choose appropriate spatial basis functions.

We begin with the thin-wire CdH-MoM formulation, namely, with the integrand on the left-hand side of the transform-domain reciprocity relation (2.6) from which we extract the transform-domain Green's function that displays the asymptotic behavior for the vanishing radius $a \downarrow 0$. Hence, making use of the integral representation for the modified Bessel function [25, eq. (9.6.23)], Eq. (A.3) at $r = a$ can be written as

$$\tilde{G}(a, p, s) = \frac{1}{2\pi} \int_{u=1}^{\infty} \exp[-s\gamma_0(p)ua] \frac{du}{(u^2 - 1)^{1/2}} \tag{14.1}$$

Time-Domain Electromagnetic Reciprocity in Antenna Modeling, First Edition. Martin Štumpf.

and recall that $\gamma_0(p) = (1/c_0^2 - p^2)^{1/2}$ with $\mathrm{Re}(\gamma_0) \geq 0$ has the meaning of the radial slowness. Substitution $q = ua$ and integration by parts then yield

$$\tilde{G}(a, p, s) = -\ln(a) \exp[-s\gamma_0(p)a]/2\pi$$

$$+ \frac{s\gamma_0(p)}{2\pi} \int_{q=a}^{\infty} \exp[-s\gamma_0(p)q] \ln[q + (q^2 - a^2)^{1/2}]\mathrm{d}q \qquad (14.2)$$

in which we next take the limit $a \downarrow 0$ and use $\exp[-s\gamma_0(p)a] = 1 + O(a)$ and $\ln[q + (q^2 - a^2)^{1/2}] = \ln(2q) + O(a^2)$. In this way, we finally end up with (see [25, eqs. (9.6.12) and (9.6.13)])

$$\tilde{G}(a, p, s) = -\{\ln[s\gamma_0(p)a/2] + \gamma\}/2\pi + O(a^2) \qquad (14.3)$$

as $a \downarrow 0$, where $\gamma = -\int_{v=0}^{\infty} \exp(-v)\mathrm{d}v$ is Euler's constant. As expected, Eq. (14.3) shows the logarithmic-type singularity of the transform-domain Green's function for the vanishing antenna's radius.

Let us next analyze the asymptotic behavior of the corresponding transform-domain term for a planar strip of a vanishing width $w \downarrow 0$ that lies in plane $z = 0$, and its axis is oriented along the x-axis. If the strip under consideration is sufficiently narrow, we may assume that the induced electric-current surface density on its planar surface has the x-oriented component only, say $\partial J^s(x, y, t)$. This component may vary, as the axial wire current $I^s(z, t)$ along a wire (see Figure 2.2a), in a piece-wire linear manner along the strip axis and time. Now a question arises what is a plausible expansion of the surface-current density in the transverse y-direction. As the Dirac-delta behavior $\delta(y)$, that would represent the current densely concentrated along $y = 0$, leads to a divergent integral representation, we will postulate a uniform distribution along the strip's width, that is $[\mathrm{H}(y + w/2) - \mathrm{H}(y - w/2)]/w$. A reason for this choice is the observation that this current distribution mimics well, in an integral sense, the behavior of the electric-current distribution that shows the inverse square-root singularity along the strip's boundaries in accordance with the edge condition [47]. Hence, with reference to the wave-slowness representation introduced in Eq. (13.10), the term corresponding to $\tilde{G}(a, p, s)$, as analyzed previously for a wire, is written as

$$(1/2\pi\mathrm{i}) \int_{\sigma=-\mathrm{i}\infty}^{\mathrm{i}\infty} \mathrm{i}_0(s\sigma w/2)\mathrm{d}\sigma/2\Gamma_0(\kappa, \sigma) \qquad (14.4)$$

for $\{s \in \mathbb{R}; s > 0\}$, $\{\kappa \in \mathbb{C}; \mathrm{Re}(\kappa) = 0\}$ and recall that $\mathrm{i}_0(x)$ denotes the modified spherical Bessel function of the first kind, $\Gamma_0(\kappa, \sigma) = [\Omega_0^2(\kappa) - \sigma^2]^{1/2}$ with $\mathrm{Re}(\Gamma_0) \geq 0$ and $\Omega_0(\kappa) = (1/c_0^2 - \kappa^2)^{1/2}$ with $\mathrm{Re}(\Omega_0) \geq 0$. The integral representation (14.4) will next be handled in a way similar to the one applied in appendix C. At first, we observe that the integrand is bounded at $\sigma = 0$, which implies that the integration contour can be indented to the right with a semicircular arc with its center at the origin and a vanishingly small radius without changing the result of integration. Consequently, the integral is written as a sum of two integrations along the indented contour, each of which being amenable to the CdH procedure. Accordingly, the (new) integration path is in virtue of Jordan's lemma and Cauchy's theorem deformed into the loops encircling the branch cuts along $\{\text{Im}(\sigma) = 0; \Omega_0(\kappa) \leq |\text{Re}(\sigma)| < \infty\}$ and around the circle of a vanishing radius with its center at the (simple) pole $\sigma = 0$. Combining the integrations along the branch cuts and adding the pole contribution, we arrive at

$$\frac{1}{2sw\Omega_0(\kappa)} - \frac{1}{sw\Omega_0(\kappa)} \frac{1}{\pi} \int_{u=1}^{\infty} \exp[-s\Omega_0(\kappa)uw/2] \frac{du}{u(u^2 - 1)^{1/2}} \qquad (14.5)$$

To cast this expression to a form similar to Eq. (14.1), we apply integration by parts and get

$$\{1 - \exp[-s\Omega_0(\kappa)w/2]\}/[2sw\Omega_0(\kappa)]$$

$$+ \frac{1}{2\pi} \int_{u=1}^{\infty} \exp[-s\Omega_0(\kappa)uw/2]\tan^{-1}[(u^2 - 1)^{-1/2}]du \qquad (14.6)$$

Apparently, the first term in Eq. (14.6) yields $1/4$ as $w \downarrow 0$, while the integral can handled via the integration by parts approach as applied to Eq. (14.1). In this way, we end up with

$$-\{\ln[s\Omega_0(\kappa)w/4] + \gamma - 1\}/2\pi + O(w^2) \qquad (14.7)$$

as $w \downarrow 0$, which shows, like Eq. (14.3) applying to a thin wire, the logarithmic singularity for the vanishing strip's width. Thus, this behavior justifies the uniform expansion of the electric-current surface density in the y-direction.

14.1 PROBLEM DESCRIPTION

The problem configuration under consideration is shown in Figure 14.1. The planar-strip antenna occupies a bounded domain in $\{-\ell/2 < x < \ell/2, -w/2 < y < w/2, z = 0\}$, where $\ell > 0$ denotes its length and $w > 0$ is its width. Again, the antenna structure is embedded in the unbounded, homogeneous, loss-free, and isotropic embedding $\mathcal{D}^\infty$ whose EM properties are described by (real-valued, positive and scalar) electric permittivity ϵ_0 and magnetic permeability μ_0 with its EM wave speed $c_0 = (\epsilon_0\mu_0)^{-1/2} > 0$. The closed surface separating the antenna from its exterior domain is denoted S_0, again.

The scattered EM wave fields, $\{\hat{E}^s, \hat{H}^s\}$, are defined as the difference between the total EM wave fields in the problem configuration and the incident EM wave fields, $\{\hat{E}^i, \hat{H}^i\}$, that represent the action of a localized voltage source in the narrow gap at $\{x_\delta - \delta/2 < x < x_\delta + \delta/2, -w/2 < y < w/2, z = 0\}$ with $\delta > 0$ and $\{-\ell/2 < x_\delta < \ell/2\}$ or/and of a uniform EM plane wave.

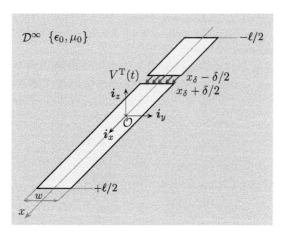

FIGURE 14.1. A planar-strip antenna excited by a voltage gap source.

14.2   PROBLEM FORMULATION

The analyzed problem is, again, formulated with the aid of the EM reciprocity theorem of the time-convolution type according to Table 2.1. This approach leads to Eq. (2.2) that is rewritten to the form reflecting the assumption that the width w is assumed to be (relatively) small with respect to the spatial support of the excitation pulse, that is (cf. Eq. (2.3))

$$\int_{x=-\ell/2}^{\ell/2} dx \int_{y=-w/2}^{w/2} \left[\hat{E}_x^B(x,y,0,s)\, \partial \hat{J}^s(x,y,s) \right.$$
$$\left. - \hat{E}_x^s(x,y,0,s)\, \partial \hat{J}^B(x,y,s) \right] dy = 0 \tag{14.8}$$

where $\partial \hat{J}^s(x,y,s)$ is the (x-component of the) (unknown) induced electric-current surface density on the antenna strip, that is, in fact, proportional to the jump of (the tangential y-component of) the scattered magnetic-field strength across the strip. Furthermore, the testing electric-field strength $\hat{E}_x^B(x,y,z,s)$ is related to the testing electric-current surface density, $\partial \hat{J}^B(x,y,s)$, according to [3, eq. (26.3-1)]

$$\hat{E}_x^B(x,y,z,s) = - s\mu_0 \hat{G}(x,y,z,s) *_{xy} \partial \hat{J}^B(x,y,s)$$
$$+ (s\epsilon_0)^{-1} \partial_x^2\, \hat{G}(x,y,z,s) *_{xy} \partial \hat{J}^B(x,y,s) \tag{14.9}$$

where $*_{xy}$ denotes the (two-dimensional) spatial convolution on the $z = 0$ plane with respect to the $\{x,y\}$-coordinates, and the support of the testing current surface density extends over the bounded domain $\{-\ell/2 < x < \ell/2, -w/2 < y < w/2, z = 0\}$. Thanks to the homogeneous, isotropic, and loss-free background, the s-domain Green's function $\hat{G}(x,y,z,s)$ is defined on the right-hand side of Eq. (2.5), again, where we use $r^2 = x^2 + y^2$ (see also [3, eq. (26.2-10)]).

Following the solution strategy introduced in section 2.2, we next find the slowness-domain counterpart of the reciprocity relation (14.8) and Eq. (14.9). Referring to the wave-slowness representation (13.10), this way leads to (cf. Eq. (2.6))

$$\left(\frac{s}{2i\pi}\right)^2 \int_{\kappa=-i\infty}^{i\infty} d\kappa \int_{\sigma=-i\infty}^{i\infty} \tilde{E}_x^B(\kappa,\sigma,0,s)\partial\tilde{J}^s(-\kappa,-\sigma,s)d\sigma$$

$$= \left(\frac{s}{2i\pi}\right)^2 \int_{\kappa=-i\infty}^{i\infty} d\kappa \int_{\sigma=-i\infty}^{i\infty} \tilde{E}_x^s(\kappa,\sigma,0,s)\partial\tilde{J}^B(-\kappa,-\sigma,s)d\sigma \qquad (14.10)$$

in which the slowness-domain testing electric-field follows from (cf. Eq. (2.7))

$$\tilde{E}_x^B(\kappa,\sigma,z,s) = -(s/\epsilon_0)\Omega_0^2(\kappa)\tilde{G}(\kappa,\sigma,z,s)\partial\tilde{J}^B(\kappa,\sigma,s) \qquad (14.11)$$

where $\Omega_0(\kappa) = (1/c_0^2 - \kappa^2)^{1/2}$ (see section F.1). The transform-domain Green's function then immediately follows upon solving Eq. (13.11) subject to the excitation condition (13.12) with $h = 0$. Accordingly, the (bounded) solution has the following form

$$\tilde{G}(\kappa,\sigma,z,s) = \exp[-s\Gamma_0(\kappa,\sigma)|z|]/2s\Gamma_0(\kappa,\sigma) \qquad (14.12)$$

where $\Gamma_0(\kappa,\sigma) = [\Omega_0^2(\kappa) - \sigma^2]^{1/2}$, in accordance with the definition given just below Eq. (13.11). The transform-domain reciprocity relation (14.10) is the point of departure for the Cagniard-DeHoop Method of Moments (CdH-MoM) described in the ensuing section.

14.3   PROBLEM SOLUTION

The problem of calculating the space-time distribution of the induced electric-current surface density on the surface of a strip is next tackled numerically. To that end, the planar surface is discretized into $N + 1$ segments of a constant length $\Delta = \ell/(N + 1) > 0$ (see Figure 14.2). The discretization points along the axis of the strip can be then specified via

$$x_n = -\ell/2 + n\,\Delta \text{ for } n = \{0, 1, \ldots, N+1\} \qquad (14.13)$$

which for $n = 0$ and $n = N + 1$ describes the end points, where the end conditions apply, that is

$$\partial\hat{J}^s(\pm\ell/2, y, s) = 0 \qquad (14.14)$$

for all $\{-w/2 < y < w/2\}$. With the uniform time grid $\{t_k = k\Delta t; \Delta t > 0, k = 1, 2, \ldots, M\}$, we may use Eqs. (2.10) and (2.11) to define the spatial and

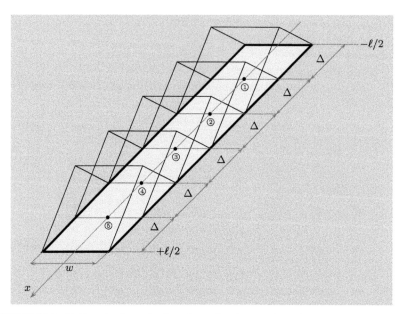

FIGURE 14.2. Uniformly discretized surface of a planar strip and the chosen spatial basis functions.

temporal piecewise-linear bases, respectively, and expand the (space-time) induced electric-current density in the following manner (see Figure 14.2)

$$\partial J^{\mathrm{s}}(x,y,t) \simeq \sum_{n=1}^{N} \sum_{k=1}^{M} j_k^{[n]} \Lambda^{[n]}(x)\Pi(y)\Lambda_k(t) \tag{14.15}$$

where $j_k^{[n]}$ are unknown coefficients (in A/m) that we seek and we used the rectangular function

$$\Pi(y) = \begin{cases} 1 & \text{for } y \in [-w/2, w/2] \\ 0 & \text{elsewhere} \end{cases} \tag{14.16}$$

in line with the observations made in the aforementioned introduction. Its transform-domain counterpart immediately follows (cf. Eq. (2.12))

$$\partial \tilde{J}^{\mathrm{s}}(\kappa,\sigma,s) \simeq \sum_{n=1}^{N} \sum_{k=1}^{M} j_k^{[n]} \tilde{\Lambda}^{[n]}(\kappa)\tilde{\Pi}(\sigma)\hat{\Lambda}_k(s) \tag{14.17}$$

The testing electric-current surface density is taken to have the "razor-type" spatial distribution and the impulsive behavior in time, that is

$$\partial J^{\mathrm{B}}(x,y,t) = \Pi^{[S]}(x)\delta(y)\delta(t) \tag{14.18}$$

for all $S = \{1, \ldots, N\}$, in which

$$\Pi^{[n]}(x) = \begin{cases} 1 \text{ for } x \in [x_n - \Delta/2, x_n + \Delta/2] \\ 0 \text{ elsewhere} \end{cases} \tag{14.19}$$

for all $n = \{1, \ldots, N\}$. Finally, under the slowness representation (13.10), we get (cf. Eq. (2.13))

$$\partial \tilde{J}^{\mathrm{B}}(\kappa, \sigma, s) = \tilde{\Pi}^{[S]}(\kappa) \tag{14.20}$$

for all $S = \{1, \ldots, N\}$, which is apparently independent of σ and s. Substitution of Eqs. (14.17) and (14.20) in the transform-domain reciprocity relation (14.10) subsequently yields a system of complex FD equations, constituents of which are amenable to handling via the CdH method. Following this procedure, we finally end up with

$$\sum_{k=1}^{m}(\boldsymbol{\mathcal{Z}}_{m-k+1} - 2\boldsymbol{\mathcal{Z}}_{m-k} + \boldsymbol{\mathcal{Z}}_{m-k-1}) \cdot \boldsymbol{J}_k = \boldsymbol{V}_m \tag{14.21}$$

where $\boldsymbol{\mathcal{Z}}_k$ is a 2-D $[N \times N]$ "impeditivity array" at $t = t_k$, $\boldsymbol{J}_k$ is an 1-D $[N \times 1]$ array of the unknown coefficients at $t = t_k$, and, finally, $\boldsymbol{V}_m$ is an 1-D $[N \times 1]$ array representing the antenna excitation at $t = t_m$. The elements of the excitation array $\boldsymbol{V}_m$ will be specified later for both the plane-wave and delta-gap excitations. Equation (14.21) is solved, again, via the step-by-step updating scheme, that is (cf. Eq. (2.15))

$$\boldsymbol{J}_m = \boldsymbol{\mathcal{Z}}_1^{-1}$$
$$\cdot \left[\boldsymbol{V}_m - \sum_{k=1}^{m-1}(\boldsymbol{\mathcal{Z}}_{m-k+1} - 2\boldsymbol{\mathcal{Z}}_{m-k} + \boldsymbol{\mathcal{Z}}_{m-k-1}) \cdot \boldsymbol{J}_k \right] \tag{14.22}$$

for all $m = \{1, \ldots, M\}$, from which the actual vector of the unknown (electric-current surface density) coefficients follows upon inverting the impeditivity matrix evaluated at $t = t_1 = \Delta t$. The elements of the impeditivity array are found using the results summarized in appendix G.

14.4 ANTENNA EXCITATION

The excitation of the strip antenna via a uniform EM plane wave and a delta-gap source is next analyzed. The main goal of the present section is to specify the elements of the excitation array $\boldsymbol{V}_m$ for the chosen excitation types.

14.4.1   Plane-Wave Excitation

If the narrow-strip antenna is irradiated by a uniform EM plane wave, the electric-field distribution over its surface can be described by (cf. Eq. (2.17))

$$\hat{E}_x^i(x, y, 0, s) = \hat{e}^i(s) \sin(\phi) \exp\{-s[\kappa_0(x + \ell/2) + \sigma_0 y]\} \qquad (14.23)$$

where $\kappa_0 = \cos(\phi) \sin(\theta)/c_0$, $\sigma_0 = \sin(\phi) \sin(\theta)/c_0$ with $\{0 \leq \phi < 2\pi\}$ and $\{0 \leq \theta \leq \pi\}$, thus applying to the plane wave polarized in parallel to the strip's plane at $z = 0$ (see Eqs. (12.41) and (12.42)). The residue theorem [30, section 3.11] can be then used to show that the slowness-domain counterpart of Eq. (14.23) can be written as

$$\tilde{E}_x^i(\kappa, \sigma, 0, s) = \hat{e}^i(s) \exp(-s\kappa_0\ell/2)$$
$$\times \ell w \sin(\phi) i_0[s(\kappa - \kappa_0)\ell/2] i_0[s(\sigma - \sigma_0)w/2] \qquad (14.24)$$

where $i_0(x)$ denotes the modified spherical Bessel function of the first kind. Assuming now that the planar strip is a perfect electrical conductor, we can make use of the (transform-domain) explicit-type boundary condition, that is

$$\tilde{E}_x^s(\kappa, \sigma, 0, s) = -\tilde{E}_x^i(\kappa, \sigma, 0, s) \qquad (14.25)$$

to specify $\tilde{E}_x^s(\kappa, \sigma, 0, s)$ on the right-hand side of the starting reciprocity relation (14.10). The transform-domain testing-current distribution (14.20) is then used to fully specify the interaction, and the resulting slowness-domain integrals are evaluated with the aid of Cauchy's formula [30, section 2.41]. This way leads to a complex-FD expression that can be readily transformed to TD, thus yielding the excitation-array elements, that is

$$V^{S]}(t) = - [\sin(\phi)/\kappa_0]e^i(t)$$
$$*_t \{H[t - \kappa_0(x_S + \ell/2 - \Delta/2)] - H[t - \kappa_0(x_S + \ell/2 + \Delta/2)]\} \qquad (14.26)$$

for all $S = \{1, \ldots, N\}$. The time convolution in Eq. (14.26) can be either calculated analytically or approximated numerically using a quadrature rule. Whenever the incident plane wave propagates along the z-axis, $\sin(\theta) = 0$ and hence $\kappa_0 = \sigma_0 = 0$. The relevant limit of Eq. (14.26) then reads

$$V^{[S]}(t) = -e^i(t)\Delta \sin(\phi) \qquad (14.27)$$

for all $S = \{1, \ldots, N\}$, again.

14.4.2 Delta-Gap Excitation

If the strip antenna is activated via the voltage pulse applied in a gap of vanishing width $\delta \downarrow 0$, then the complex-FD incident field can be described by

$$\hat{E}_x^i(x, y, 0, s) = \hat{V}^T(s)\delta(x - x_\delta)\Pi(y) \tag{14.28}$$

where we used $\Pi(y)$ defined by Eq. (14.16), $\{-\ell/2 < x_\delta < \ell/2\}$ denotes the center of the gap and $\hat{V}^T(s)$ is the complex-FD counterpart of the excitation voltage pulse (see Figure 14.1). The residue theorem [30, section 3.11] can be then applied to show that the slowness-domain counterpart of Eq. (14.28) has the form

$$\tilde{E}_x^i(\kappa, \sigma, 0, s) = \hat{V}^T(s)w \exp(s\kappa x_\delta)i_0(s\sigma w/2) \tag{14.29}$$

Again, employing the explicit-type boundary condition (14.25) with Eqs. (14.29) and (14.20), the right-hand side of Eq. (14.10) can be evaluated via Cauchy's formula [30, section. 2.41]. Transforming the result to the TD, we end up with the following expression specifying the elements of the excitation array for the delta-gap voltage excitation, that is (cf. Eqs. (2.25) and (2.26))

$$V^{[S]}(t) = -V^T(t)[H(x_\delta + \Delta/2 - x_S) - H(x_\delta - \Delta/2 - x_S)] \tag{14.30}$$

for all $S = \{1, \ldots, N\}$.

14.5 EXTENSION TO A WIDE-STRIP ANTENNA

If the width of a strip antenna is no longer narrow with respect to the spatial support of the excitation pulse, the variation of the induced electric current along the y-direction must be properly accounted for in the solution procedure. It is next shown that this can be done via a straightforward extension of the methodology concerning a narrow strip.

Keeping in mind the vectorial distribution of the electric-current surface density, the starting reciprocity relation (14.8) is generalized accordingly (cf. Eq. (2.2))

$$\int_{x \in S} \left[\hat{E}^B(x, y, 0, s) \cdot \partial \hat{J}^s(x, y, s) \right.$$
$$\left. - \hat{E}^s(x, y, 0, s) \cdot \partial \hat{J}^B(x, y, s) \right] dA = 0 \tag{14.31}$$

where S denotes the bounded surface that lies in $z = 0$, and $\partial \hat{J}^{s,B} = \partial \hat{J}_x^{s,B} i_x + \partial \hat{J}_y^{s,B} i_y$ are the induced and testing electric-current surface densities, respectively, whose support is $\mathcal{S} \subset \mathbb{R}^2$. The corresponding slowness-domain reciprocity relation then has the following form (cf. Eq. (14.10))

$$
\left(\frac{s}{2i\pi}\right)^2 \int_{\kappa=-i\infty}^{i\infty} d\kappa \int_{\sigma=-i\infty}^{i\infty} \left[\tilde{E}_x^{\text{B}}(\kappa,\sigma,0,s)\partial\tilde{J}_x^{\text{s}}(-\kappa,-\sigma,s)\right.
$$
$$
\left. + \tilde{E}_y^{\text{B}}(\kappa,\sigma,0,s)\partial\tilde{J}_y^{\text{s}}(-\kappa,-\sigma,s)\right] d\sigma
$$
$$
= \left(\frac{s}{2i\pi}\right)^2 \int_{\kappa=-i\infty}^{i\infty} d\kappa \int_{\sigma=-i\infty}^{i\infty} \left[\tilde{E}_x^{\text{s}}(\kappa,\sigma,0,s)\partial\tilde{J}_x^{\text{B}}(-\kappa,-\sigma,s)\right.
$$
$$
\left. + \tilde{E}_y^{\text{s}}(\kappa,\sigma,0,s)\partial\tilde{J}_y^{\text{B}}(-\kappa,-\sigma,s)\right] d\sigma \tag{14.32}
$$

in which the (transform-domain) testing electric-field strength is related to its source via (cf. Eq. (14.11))

$$
\tilde{E}_x^{\text{B}}(\kappa,\sigma,z,s) = -(s/\epsilon_0)\Omega_0^2(\kappa)\tilde{G}(\kappa,\sigma,z,s)\partial\tilde{J}_x^{\text{B}}(\kappa,\sigma,s)
$$
$$
+ (s/\epsilon_0)\kappa\sigma\,\tilde{G}(\kappa,\sigma,z,s)\partial\tilde{J}_y^{\text{B}}(\kappa,\sigma,s) \tag{14.33}
$$
$$
\tilde{E}_y^{\text{B}}(\kappa,\sigma,z,s) = -(s/\epsilon_0)\Omega_0^2(\sigma)\tilde{G}(\kappa,\sigma,z,s)\partial\tilde{J}_y^{\text{B}}(\kappa,\sigma,s)
$$
$$
+ (s/\epsilon_0)\sigma\kappa\,\tilde{G}(\kappa,\sigma,z,s)\partial\tilde{J}_x^{\text{B}}(\kappa,\sigma,s) \tag{14.34}
$$

where $\Omega_0(\kappa) = (1/c_0^2 - \kappa^2)^{1/2}$, again. To solve the slowness-domain reciprocity relation numerically, the antenna's surface is divided into rectangular cells of identical dimensions Δ_x and Δ_y along the corresponding directions. It is noted that the uniform grid is not mandatory, but it greatly simplifies the numerical solution and hence its code implementation. Following the line of reasoning pursued in section 14.3, the induced electric-current surface density is subsequently expanded in a piecewise linear manner both in space and time, that is (cf. Eq. (14.15))

$$
\partial J_x^{\text{s}}(x,y,t) \simeq \frac{1}{\Delta_y}\sum_{u=1}^{U}\sum_{k=1}^{M} i_k^{[u]}\Lambda^{[u]}(x)\Pi^{[u]}(y)\Lambda_k(t) \tag{14.35}
$$

where $i_k^{[u]}$ are unknown coefficients (in A) pertaining to the x-component of the electric current at the u-th spatial node and at $t = t_k = k\Delta t > 0$. An example of the x-directed "roof-top function" $\Lambda^{[u]}(x)$ for $u = 3$ is given in Figure 14.3 (see also Eq. (2.10)) and $\Pi^{[u]}(y)$ is the corresponding rectangular function that is defined in a similar way as Eq. (14.19), that is

$$
\Pi^{[u]}(y) = \begin{cases} 1 \text{ for } y \in [y_u - \Delta_y/2, y_u + \Delta_y/2] \\ 0 \text{ elsewhere} \end{cases} \tag{14.36}
$$

for all $u = \{1,\ldots,U\}$, and, finally, $\Lambda_k(t)$ was defined in Eq. (2.11). Adopting the similar strategy for the y-directed induced current, we may write

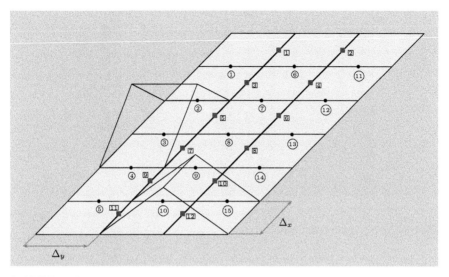

FIGURE 14.3. Uniformly discretized surface of a planar antenna, and examples of x-directed and y-directed spatial basis functions.

$$\partial J_y^s(x,y,t) \simeq \frac{1}{\Delta_x} \sum_{v=1}^{V} \sum_{k=1}^{M} i_k^{[v]} \Lambda^{[v]}(y) \Pi^{[v]}(x) \Lambda_k(t) \tag{14.37}$$

in which $i_k^{[v]}$ denote coefficients (in A) pertaining to the y-component of the electric current at the v-th spatial node and at $t = t_k = k\Delta t > 0$, again. For the sake of illustration, the y-directed basis function $\Lambda^{[v]}(y)$ for $v = 12$ is depicted in Figure 14.3. It remains to specify the testing electric-current distribution. To that end, we may use the "razor-type" testing current (see Eq. (14.18)), again, and take

$$\partial J_x^B(x,y,t) = \Pi^{[P]}(x)\delta(y - y_P)\delta(t) \tag{14.38}$$

$$\partial J_y^B(x,y,t) = \Pi^{[Q]}(y)\delta(x - x_Q)\delta(t) \tag{14.39}$$

for all $P = \{1, \ldots, U\}$ and $Q = \{1, \ldots, V\}$, where the meaning of rectangular functions $\Pi^{[P]}(x)$ and $\Pi^{[Q]}(y)$ is clear from Eq. (14.36). Subsequently, the transform-domain counterparts of Eqs. (14.35), (14.37)–(14.39) are substituted in the reciprocity relation (14.32), which yields a system of equations in the complex-FD that can be cast into the matrix form depicted as in Figure 14.4. Symbolically, the 1-D arrays $\hat{I}_x^{[u]}$ and $\hat{I}_y^{[v]}$ consist of $[U \times 1]$ and $[V \times 1]$ elements, respectively, representing the (unknown) coefficients of the x-directed and y-directed induced currents. Furthermore, the 1-D voltage arrays $\hat{V}_x^{[P]}$ and $\hat{V}_y^{[Q]}$ of dimensions $[U \times 1]$ and $[V \times 1]$, respectively, are representatives of the (weighted)

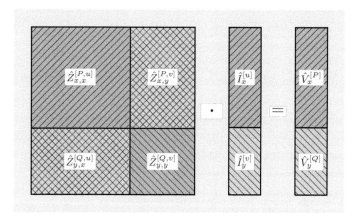

FIGURE 14.4. Symbolical impedance-matrix description of a planar-antenna problem.

x- and y-components of the incident electric-field strength on the conductive antenna surface $\mathcal{S}$. The electric-current and excitation voltage 1-D arrays are mutually interrelated through the impedance array, whose elements are next closely described. We begin with the $[U \times U]$ partial impedance matrix $\hat{Z}_{x,x}^{[P,u]}$, whose elements are, in fact, closely related to the ones characterizing a narrow strip oriented along the x-axis (see Figure 14.1). Hence, referring to Eq. (G.1), we may write

$$\hat{Z}_{x,x}^{[P,u]}(s) = [\zeta_0/c_0 \Delta t \Delta_x \Delta_y]$$
$$\times \Big[\hat{J}(x_P - x_u + 3\Delta_x/2, y_P - y_u + \Delta_y/2, s)$$
$$- \hat{J}(x_P - x_u + 3\Delta_x/2, y_P - y_u - \Delta_y/2, s)$$
$$- 3\hat{J}(x_P - x_u + \Delta_x/2, y_P - y_u + \Delta_y/2, s)$$
$$+ 3\hat{J}(x_P - x_u + \Delta_x/2, y_P - y_u - \Delta_y/2, s)$$
$$+ 3\hat{J}(x_P - x_u - \Delta_x/2, y_P - y_u + \Delta_y/2, s)$$
$$- 3\hat{J}(x_P - x_u - \Delta_x/2, y_P - y_u - \Delta_y/2, s)$$
$$- \hat{J}(x_P - x_u - 3\Delta_x/2, y_P - y_u + \Delta_y/2, s)$$
$$+ \hat{J}(x_P - x_u - 3\Delta_x/2, y_P - y_u - \Delta_y/2, s) \Big] \qquad (14.40)$$

for all $P = \{1, \ldots, U\}$ and $u = \{1, \ldots, U\}$, where $\hat{J}(x, y, s)$ is given in appendix G. In a similar way, we may represent the components of the square $[V \times V]$ partial impedance matrix $\hat{Z}_{y,y}^{[Q,v]}(s)$, that is

$$\hat{Z}_{y,y}^{[Q,v]}(s) = [\zeta_0/c_0\Delta t\Delta_y\Delta_x]$$

$$\times \left[\hat{J}(y_Q - y_v + 3\Delta_y/2, x_Q - x_v + \Delta_x/2, s)\right.$$

$$- \hat{J}(y_Q - y_v + 3\Delta_y/2, x_Q - x_v - \Delta_x/2, s)$$

$$- 3\hat{J}(y_Q - y_v + \Delta_y/2, x_Q - x_v + \Delta_x/2, s)$$

$$+ 3\hat{J}(y_Q - y_v + \Delta_y/2, x_Q - x_v - \Delta_x/2, s)$$

$$+ 3\hat{J}(y_Q - y_v - \Delta_y/2, x_Q - x_v + \Delta_x/2, s)$$

$$- 3\hat{J}(y_Q - y_v - \Delta_y/2, x_Q - x_v - \Delta_x/2, s)$$

$$- \hat{J}(y_Q - y_v - 3\Delta_y/2, x_Q - x_v + \Delta_x/2, s)$$

$$\left.+ \hat{J}(y_Q - y_v - 3\Delta_y/2, x_Q - x_v - \Delta_x/2, s)\right] \tag{14.41}$$

for all $Q = \{1, \ldots, V\}$ and $v = \{1, \ldots, V\}$. The principal difference with respect to the solution concerning a narrow strip (see section 14.1) is the coupling between the mutually orthogonal "currents" and "voltages," namely, $\hat{I}_x^{[u]} \to \hat{V}_y^{[Q]}$ and $\hat{I}_y^{[v]} \to \hat{V}_x^{[P]}$. The latter is effectuated via submatrices $\hat{Z}_{y,x}^{[Q,u]}$ and $\hat{Z}_{x,y}^{[P,v]}$, respectively, whose elements next follow

$$\hat{Z}_{y,x}^{[Q,u]}(s) = \left[\zeta_0/c_0\Delta t\Delta_x\Delta_y\right]$$

$$\times \left[\hat{I}(x_Q - x_u + \Delta_x, y_Q - y_u + \Delta_y, s) - 2\hat{I}(x_Q - x_u + \Delta_x, y_Q - y_u, s)\right.$$

$$+ \hat{I}(x_Q - x_u + \Delta_x, y_Q - y_u - \Delta_y, s) - 2\hat{I}(x_Q - x_u, y_Q - y_u + \Delta_y, s)$$

$$+ 4\hat{I}(x_Q - x_u, y_Q - y_u, s) - 2\hat{I}(x_Q - x_u, y_Q - y_u - \Delta_y, s)$$

$$+ \hat{I}(x_Q - x_u - \Delta_x, y_Q - y_u + \Delta_y, s) - 2\hat{I}(x_Q - x_u - \Delta_x, y_Q - y_u, s)$$

$$\left.+ \hat{I}(x_Q - x_u - \Delta_x, y_Q - y_u - \Delta_y, s)\right] \tag{14.42}$$

for all $Q = \{1, \ldots, V\}$ and $u = \{1, \ldots, U\}$, and similarly,

$$\hat{Z}_{x,y}^{[P,v]}(s) = [\zeta_0/c_0\Delta t\Delta_y\Delta_x]$$

$$\times \left[\hat{I}(y_P - y_v + \Delta_y, x_P - x_v + \Delta_x, s) - 2\hat{I}(y_P - y_v + \Delta_y, x_P - x_v, s)\right.$$

$$+ \hat{I}(y_P - y_v + \Delta_y, x_P - x_v - \Delta_x, s) - 2\hat{I}(y_P - y_v, x_P - x_v + \Delta_x, s)$$

$$+ 4\hat{I}(y_P - y_v, x_P - x_v, s) - 2\hat{I}(y_P - y_v, x_P - x_v - \Delta_x, s)$$

$$+ \hat{I}(y_P - y_v - \Delta_y, x_P - x_v + \Delta_x, s) - 2\hat{I}(y_P - y_v - \Delta_y, x_P - x_v, s)$$

$$\left.+ \hat{I}(y_P - y_v - \Delta_y, x_P - x_v - \Delta_x, s)\right] \tag{14.43}$$

for all $P = \{1, \ldots, U\}$ and $v = \{1, \ldots, V\}$. In Eqs. (14.42) and (14.43), we used $\hat{I}(x, y, s)$ that can be represented via (cf. Eq. (G.3))

$$\hat{I}(x, y, s) = -\frac{c_0^2}{8\pi^2 s^2} \int_{\kappa \in \mathbb{K}_0} \frac{\exp(s\kappa x)}{s\kappa} d\kappa$$

$$\times \int_{\sigma \in \mathbb{S}_0} \frac{\exp(s\sigma y)}{s\sigma} \frac{d\sigma}{\Gamma_0(\kappa, \sigma)} \tag{14.44}$$

for $\{x \in \mathbb{R}; x \neq 0\}$, $\{y \in \mathbb{R}; y \neq 0\}$, and $\{s \in \mathbb{R}; s > 0\}$, where the (indented) integration paths $\mathbb{K}_0$ and $\mathbb{S}_0$ are shown in Figure G.1.

The system of equations is subsequently transformed back to the TD. As the transformation of $\hat{J}(x, y, s)$ has been carried out in appendix G, it remains to determine the TD counterpart of $\hat{I}(x, y, s)$. This can be done using the CdH technique along the lines closely described in appendix G. This strategy finally yields

$$I(x, y, t) = c_0^3 t^3 \mathrm{H}(x)\mathrm{H}(y)\mathrm{H}(t)/12$$

$$+ \frac{\mathrm{sgn}(y)\mathrm{H}(x)}{4\pi} \left\{ |y| \left(c_0^2 t^2 + \frac{y^2}{6} \right) \cosh^{-1}\left(\frac{c_0 t}{|y|} \right) \right.$$

$$- \frac{c_0^3 t^3}{3} \tan^{-1}\left[\left(\frac{c_0^2 t^2}{y^2} - 1 \right)^{1/2} \right]$$

$$\left. - \frac{5}{6} c_0 t y^2 \left(\frac{c_0^2 t^2}{y^2} - 1 \right)^{1/2} \right\} \mathrm{H}(c_0 t - |y|)$$

$$+ \frac{\mathrm{sgn}(x)\mathrm{H}(y)}{4\pi} \left\{ |x| \left(c_0^2 t^2 + \frac{x^2}{6} \right) \cosh^{-1}\left(\frac{c_0 t}{|x|} \right) \right.$$

$$- \frac{c_0^3 t^3}{3} \tan^{-1}\left[\left(\frac{c_0^2 t^2}{x^2} - 1 \right)^{1/2} \right]$$

$$\left. - \frac{5}{6} c_0 t x^2 \left(\frac{c_0^2 t^2}{x^2} - 1 \right)^{1/2} \right\} \mathrm{H}(c_0 t - |x|)$$

$$+ \frac{1}{12\pi} \int_{v=r}^{c_0 t} (c_0 t - v)^3 \overline{P}(x, y, v) dv \tag{14.45}$$

in which $r = (x^2 + y^2)^{1/2} > 0$ and (cf. Eq. (G.20))

$$\overline{P}(x, y, c_0 \tau) = [\mathrm{sgn}(x)\mathrm{sgn}(y)/2c_0\tau]$$

$$\times \left[(c_0^2 \tau^2/x^2 - 1)^{-1/2} + (c_0^2 \tau^2/y^2 - 1)^{-1/2} \right] \tag{14.46}$$

The convolution integral in Eq. (14.45) can be handled numerically with the aid of the recursive convolution method (see appendix H) or analytically. The latter way results in

$$\frac{1}{12\pi}\int_{v=r}^{c_0 t}(c_0 t - v)^3 \overline{P}(x,y,v)\mathrm{d}v = [\mathrm{sgn}(x)\mathrm{sgn}(y)/24\pi]$$
$$\times [N(x,y,t) + N(y,x,t)] \qquad (14.47)$$

where

$$N(x,y,t) = c_0^3 t^3 \left\{ \tan^{-1}[(c_0^2 t^2 - x^2)^{1/2}/|x|] - \tan^{-1}(|y|/|x|) \right\}$$
$$- 3|x|(c_0^2 t^2 + x^2/6)\ln\{[c_0 t + (c_0^2 t^2 - x^2)^{1/2}]/(r+|y|)\}$$
$$+ 5|x|c_0 t(c_0^2 t^2 - x^2)^{1/2}/2 - 3|x||y|(c_0 t - r/6) \qquad (14.48)$$

With the TD functions $J(x,y,t)$ and $I(x,y,t)$ at our disposal, the complete TD impedance matrix immediately follows. In this manner, the starting reciprocity relation (14.32) can be cast into the form of Eq. (2.14), whose solution leads to the electric-current coefficients $i_k^{[u]}$ and $i_k^{[v]}$ for all $u = \{1, \ldots, U\}$, $v = \{1, \ldots, V\}$, and $k = \{1, \ldots, M\}$, thus determining the approximate space-time distribution of the induced current on the antenna's surface (cf. Eqs. (14.35) and (14.37)).

In conclusion, we plot the electric-current surface-density distribution on a PEC sheet of dimensions $\ell = 100$ mm and $w = 50$ mm induced by a uniform EM plane wave whose propagation vector $\boldsymbol{\beta} = -\boldsymbol{i}_z$ is perpendicular to the PEC surface and whose polarization vector is defined via $\boldsymbol{\alpha} = \boldsymbol{i}_x \sin(\phi) - \boldsymbol{i}_y \cos(\phi)$. The corresponding excitation-array elements can be then found from (see Eq. (14.27) and Figure 14.4)

$$V_x^{[P]}(t) = -e^{\mathrm{i}}(t)\Delta_x \sin(\phi) \qquad (14.49)$$

$$V_y^{[Q]}(t) = e^{\mathrm{i}}(t)\Delta_y \cos(\phi) \qquad (14.50)$$

for all $P = \{1, \ldots, U\}$ and $Q = \{1, \ldots, V\}$. Figure 14.5 shows the resulting electric-current surface distributions for $\phi = \pi/4$ and the triangular plane-wave signature $e^{\mathrm{i}}(t)$ as shown in Figure 7.2b. The reference of the plane wave was chosen such that it hits the screen at $t = 0$.

ILLUSTRATIVE NUMERICAL EXAMPLE

- Make use of the concept of equivalent radius [17] to validate the computational methodology described in section 14.3 using the thin-wire CdH-MoM formulation introduced in section 2.3.

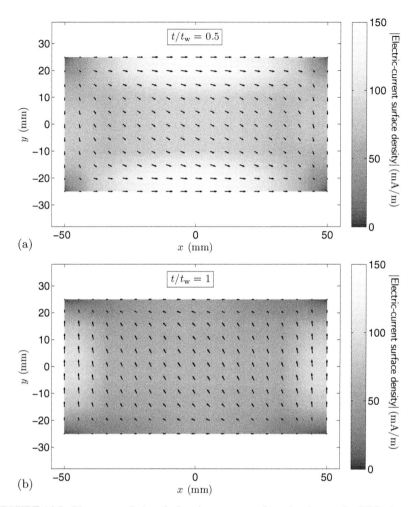

FIGURE 14.5. Plane-wave induced electric-current surface density on the PEC sheet at (a) $t/t_w = 0.5$; (b) $t/t_w = 1.0$.

Solution: According to the analysis presented in Ref. [17], the equivalent radius of a narrow PEC strip is equal to one-fourth of its width. This result has been found upon postulating the electric-current surface-density distribution that, in virtue of the relevant edge condition, exhibits the inverse square-root singularities at $y = \pm w/2$. From the results of Ref. [47], it follows that for the uniform distribution along the y-direction (see Eq. (14.15)), the equivalent radius is no longer $a = w/4$, but must be modified according to

$$a = w \exp(-3/2) \simeq 0.2231w \qquad (14.51)$$

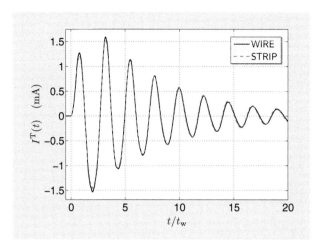

FIGURE 14.6. Electric-current self-response of the thin-wire antenna (WIRE) and the (equivalent) narrow-strip antenna (STRIP).

As a validation example, we shall largely adopt the wire antenna considered at the end of chapter 7. Hence, we shall first analyze the thin-wire antenna of length $\ell = 0.10$ m whose radius is $a = 0.10$ mm. The antenna is now excited at $z_\delta = \ell/5$. The equivalent narrow-strip antenna has the same length, but its width follows from Eq. (14.51) as $w = 0.10 \cdot \exp(3/2)$ mm $\simeq 0.4482$ mm. The both antenna structures are excited via the voltage pulse specified by Eq. (7.4), in which we take $V_m = 1.0$ V and $c_0 t_w = 1.0\ \ell$, again (see Figure 7.2a). The electric-current (self-)responses of the thin-wire and equivalent-strip antennas as observed in the time window $\{0 \leq t/t_w < 20\}$ are shown in Figure 14.6. Despite the differences in the computational models and their numerical handling, the resulting electric-current pulse shapes do converge to each other, thus demonstrating the equivalence and validity of the introduced numerical procedures.

CHAPTER 15

INCORPORATING STRIP-ANTENNA LOSSES

In section 14.4, we have incorporated the antenna excitation via the explicit-type boundary condition applying to the axial component of the electric-field strength on the PEC surface of a strip antenna. If the planar strip is no longer EM-impenetrable, the concept of high-contrast, thin-sheet cross-boundary conditions [18] lends itself to incorporate the effect of a finite strip conductivity and permittivity. Accordingly, the boundary condition on the surface of a planar-strip antenna is modified with the aid of the high-contrast thin-sheet jump condition derived in appendix I, that is (cf. Eq. (4.1))

$$\hat{E}^{\mathrm{s}}_x(x, y, 0, s) = -\hat{E}^{\mathrm{i}}_x(x, y, 0, s) + \hat{Z}^{\mathrm{L}}(s)\partial\hat{J}^{\mathrm{s}}(x, y, s) \tag{15.1}$$

for all $\{-\ell/2 < x < \ell/2, -w/2 < y < w/2\}$, with $\hat{Z}^{\mathrm{L}}(s) = 1/[\hat{G}^{\mathrm{L}}(s) + s\hat{C}^{\mathrm{L}}(s)]$, where $\hat{G}^{\mathrm{L}}(s)$ and $\hat{C}^{\mathrm{L}}(s)$ are the conductance and capacitance parameters of the thin-strip layer, respectively (see Eqs. (I.8) and (I.9)). Apparently, if either $\hat{G}^{\mathrm{L}}(s) \to \infty$ or $\hat{C}^{\mathrm{L}}(s) \to \infty$, then $\hat{Z}^{\mathrm{L}}(s) \to 0$, thus arriving at the explicit-type boundary condition that applies to the PEC surface (see Eq. (14.25)). Following next the strategy described in chapter 4, the slowness-domain counterpart of Eq. (15.1) is substituted in the right-hand side of the starting reciprocity relation (14.10), which results in (cf. Eq. (4.2))

$$\left(\frac{s}{2\mathrm{i}\pi}\right)^2 \int_{\kappa=-\mathrm{i}\infty}^{\mathrm{i}\infty} \mathrm{d}\kappa \int_{\sigma=-\mathrm{i}\infty}^{\mathrm{i}\infty} \tilde{E}^{\mathrm{s}}_x(\kappa, \sigma, 0, s)\partial\tilde{J}^{\mathrm{B}}(-\kappa, -\sigma, s)\mathrm{d}\sigma$$

$$= -\left(\frac{s}{2\mathrm{i}\pi}\right)^2 \int_{\kappa=-\mathrm{i}\infty}^{\mathrm{i}\infty} \mathrm{d}\kappa \int_{\sigma=-\mathrm{i}\infty}^{\mathrm{i}\infty} \tilde{E}^{\mathrm{i}}_x(\kappa, \sigma, 0, s)\partial\tilde{J}^{\mathrm{B}}(-\kappa, -\sigma, s)\mathrm{d}\sigma$$

$$+ \hat{Z}^{\mathrm{L}}(s)\left(\frac{s}{2\mathrm{i}\pi}\right)^2 \int_{\kappa=-\mathrm{i}\infty}^{\mathrm{i}\infty} \mathrm{d}\kappa \int_{\sigma=-\mathrm{i}\infty}^{\mathrm{i}\infty} \partial\tilde{J}^{\mathrm{s}}(\kappa, \sigma, s)\partial\tilde{J}^{\mathrm{B}}(-\kappa, -\sigma, s)\mathrm{d}\sigma \tag{15.2}$$

Time-Domain Electromagnetic Reciprocity in Antenna Modeling, First Edition. Martin Štumpf.
© 2020 by The Institute of Electrical and Electronics Engineers, Inc. Published 2020 by John Wiley & Sons, Inc.

The first term on the right-hand side of Eq. (15.2) has been in section 14.4 transformed to the TD for the plane-wave and voltage-gap excitations. The handling of the second interaction term is the subject of the ensuing section.

15.1 MODIFICATION OF THE IMPEDITIVITY MATRIX

Substituting Eqs. (14.17) and (14.20) in the second interaction on the right-hand side of Eq. (15.2), it is straightforward to show that the impact of finite strip conductivity and permittivity can be accounted for via an additional impeditivity matrix, whose elements follow from (cf. Eq. (G.1))

$$
\hat{\mathcal{R}}^{[S,n]}(s) = [\hat{Z}^{\mathrm{L}}(s)/c_0\Delta t\Delta][\hat{I}(x_n - x_S + 3\Delta/2, w/2, s)
$$
$$
- 3\hat{I}(x_n - x_S + \Delta/2, w/2, s) + 3\hat{I}(x_n - x_S - \Delta/2, w/2, s)
$$
$$
- \hat{I}(x_n - x_S - 3\Delta/2, w/2, s)] \tag{15.3}
$$

for all $S = \{1, \ldots, N\}$ and $n = \{1, \ldots, N\}$, where function $\hat{I}(x, y, s)$ is given by

$$
\hat{I}(x, y, s) = c_0\left(\frac{1}{2\mathrm{i}\pi}\right)^2 \int_{\kappa\in\mathbb{K}_0} \frac{\exp(s\kappa x)}{s^3\kappa^3}\,\mathrm{d}\kappa \int_{\sigma\in\mathbb{S}_0} \frac{\exp(s\sigma y)}{s\sigma}\,\mathrm{d}\sigma \tag{15.4}
$$

for $x \in \mathbb{R}$ and $y \in \mathbb{R}$, and the (indented) integration contours $\mathbb{K}_0$ and $\mathbb{S}_0$ are depicted in Figure G.1. Upon noting that the integrands in Eq. (15.4) do show only pole singularities at the origins of the complex κ- and σ-planes, the integrations can be readily handled via Cauchy's formula [30, section 2.41]. Using the formula, we end up with

$$
\hat{I}(x, y, s) = (c_0 x^2/2s^2)\mathrm{H}(x)\mathrm{H}(y) \tag{15.5}
$$

for all $x \in \mathbb{R}$ and $y \in \mathbb{R}$. With the closed-form expression for $\hat{I}(x, y, s)$ at our disposal, we may next proceed with the transformation of Eq. (15.3) back to the TD. Since the result of this step depends on the form of the layer's impedance $\hat{Z}^{\mathrm{L}}(s)$, it is convenient to express (the TD elements of) the impeditivity matrix as

$$
\mathcal{R}^{[S,n]}(t) = \mathcal{R}(x_n - x_S + 3\Delta/2, t) - 3\mathcal{R}(x_n - x_S + \Delta/2, t)
$$
$$
+ 3\mathcal{R}(x_n - x_S - \Delta/2, t) - \mathcal{R}(x_n - x_S - 3\Delta/2, t) \tag{15.6}
$$

for all $S = \{1, \ldots, N\}$ and $n = \{1, \ldots, N\}$, where $\mathcal{R}(x, t)$ is the TD counterpart of

$$
\hat{\mathcal{R}}(x, s) = [\hat{Z}^{\mathrm{L}}(s)/c_0\Delta t\Delta]\hat{I}(x, w/2, s) \tag{15.7}
$$

Once the elements of the impeditivity matrix $\underline{\mathcal{R}}$ are evaluated, the updating scheme (14.22) is replaced with

$$J_m = \overline{\underline{Z}}_1^{-1}$$

$$\cdot \left[V_m - \sum_{k=1}^{m-1} (\overline{\underline{Z}}_{m-k+1} - 2\overline{\underline{Z}}_{m-k} + \overline{\underline{Z}}_{m-k-1}) \cdot J_k \right] \tag{15.8}$$

in which

$$\overline{\underline{Z}} = \underline{Z} - \underline{\mathcal{R}} \tag{15.9}$$

is the modified impeditivity matrix including losses in the strip antenna, and the elements of $\underline{Z}$ are specified in appendix G. Explicit expressions for $\mathcal{R}(x, t)$ concerning selected important cases will be discussed in the following parts. More complex dispersion models can be readily incorporated along the same lines.

15.1.1   Strip with Conductive Properties

If the strip antenna is made of a (relatively highly) conductive material showing no contrast in other EM constitutive parameters with respect to the embedding, then Eq. (15.7) has the form

$$\hat{\mathcal{R}}(x, s) = \frac{1}{2G^{\mathrm{L}}} \frac{x^2}{c_0 \Delta t \Delta} \frac{c_0}{s^2} \mathrm{H}(x) \tag{15.10}$$

where $G^{\mathrm{L}} = \delta \sigma = O(1)$ as $\delta \downarrow 0$, for a homogeneous conductive strip of conductivity σ (see Eq. (I.8)). The TD counterpart of Eq. (15.10) then reads

$$\mathcal{R}(x, t) = \frac{c_0 t}{2G^{\mathrm{L}}} \frac{x^2}{c_0 \Delta t \Delta} \mathrm{H}(x) \mathrm{H}(t) \tag{15.11}$$

The use of the latter expression in Eq. (15.6) finally yields the desired TD impeditivity matrix $\underline{\mathcal{R}}$.

15.1.2   Strip with Dielectric Properties

For the strip antenna composed of a (relatively) high-permittivity dielectric material, Eq. (15.7) has the following form

$$\hat{\mathcal{R}}(x, s) = \frac{1}{2C^{\mathrm{L}}} \frac{x^2}{c_0 \Delta t \Delta} \frac{c_0}{s^3} \mathrm{H}(x) \tag{15.12}$$

where $C^{\mathrm{L}} = \delta \epsilon = O(1)$ as $\delta \downarrow 0$, for a homogeneous dielectric strip of permittivity ϵ (see Eq. (I.9)). Carrying out the transformation to the TD, we obtain at once

$$\mathcal{R}(x, t) = \frac{c_0^2 t^2}{4c_0 C^{\mathrm{L}}} \frac{x^2}{c_0 \Delta t \Delta} \mathrm{H}(x) \mathrm{H}(t) \tag{15.13}$$

Again, substitution of Eq. (15.13) in Eq. (15.6) leads to $\mathcal{R}$, thereby accounting for the effect of the finite dielectric constant.

15.1.3 Strip with Conductive and Dielectric Properties

In case that the strip antenna shows a high EM contrast in both its conductive and dielectric properties, Eq. (15.7) can be written as

$$\hat{\mathcal{R}}(x, s) = \frac{1}{2C^{\mathrm{L}}} \frac{x^2}{c_0 \Delta t \Delta} \frac{c_0}{s^2} \frac{1}{s + G^{\mathrm{L}}/C^{\mathrm{L}}} \mathrm{H}(x) \tag{15.14}$$

where, again, $C^{\mathrm{L}} = \delta\epsilon = O(1)$ and $G^{\mathrm{L}} = \delta\sigma = O(1)$ as $\delta \downarrow 0$ for a homogeneous strip (see Eqs. (I.8) and (I.9)). Letting, for brevity, $\alpha = G^{\mathrm{L}}/C^{\mathrm{L}}$, the TD counterpart of Eq. (15.14) can be written as

$$\mathcal{R}(x, t) = \frac{c_0^2 t^2}{2c_0 C^{\mathrm{L}}} \left[\frac{1}{\alpha t} - \frac{1 - \exp(-\alpha t)}{\alpha^2 t^2} \right] \frac{x^2}{c_0 \Delta t \Delta} \mathrm{H}(x)\mathrm{H}(t) \tag{15.15}$$

which is finally substituted in Eq. (15.6), thus incorporating the effect of conductive and dielectric properties.

15.1.4 Strip with Drude-Type Dispersion

The plasmonic behavior of conduction electrons in the strip antenna can be described via the Drude-type conduction relaxation function of an isotropic metal [3, section 19.5]

$$\hat{\sigma}(s) = \epsilon_0 \omega_{\mathrm{pe}}^2/(s + \nu_{\mathrm{c}}) \tag{15.16}$$

where ω_{pe} is the electron plasma angular frequency and ν_{c} denotes the collision frequency. Considering a homogeneous strip, again, Eq. (15.7) has the following form

$$\hat{\mathcal{R}}(x, s) = \frac{c_0}{2\delta\epsilon_0\omega_{\mathrm{pe}}^2} \frac{x^2}{c_0 \Delta t \Delta} \frac{s + \nu_{\mathrm{c}}}{s^2} \mathrm{H}(x) \tag{15.17}$$

The inverse Laplace transformation can be, again, carried out at once, which results in

$$\mathcal{R}(x, t) = \frac{c_0}{2\delta\epsilon_0\omega_{\mathrm{pe}}^2} \frac{x^2}{c_0 \Delta t \Delta} (1 + \nu_{\mathrm{c}} t)\mathrm{H}(x)\mathrm{H}(t) \tag{15.18}$$

The plasmonic behavior of a thin-strip antenna is finally included in upon substituting the latter expression in Eq. (15.6) and carrying out the calculations according to Eq. (15.8) with (15.9).

CHAPTER 16

CONNECTING A LUMPED ELEMENT TO THE STRIP ANTENNA

To properly incorporate a lumped element into the CdH-MoM as formulated in chapter 14, we may adopt the strategy previously applied to (loaded) thin-wire antennas (see chapter 5). Along these lines, we first modify the boundary condition (15.1) by assuming a localized load uniformly distributed over the width of the strip at $x = x_\zeta$, that is

$$\hat{E}_x^s(x, y, 0, s) = -\hat{E}_x^i(x, y, 0, s) + w\hat{\zeta}(s)\delta(x - x_\zeta)\Pi(y)\partial \hat{J}^s(x, y, s) \qquad (16.1)$$

for all $\{-\ell/2 < x < \ell/2, -w/2 < y < w/2\}$ with $\{-\ell/2 < x_\zeta < \ell/2\}$, where $\hat{\zeta}(s)$, again, denotes the impedance of the lumped element, and $\Pi(y)$ is defined by Eq. (14.16). In the next step, the slowness-domain counterpart is substituted in (the right-hand side of) Eq. (14.10). This yields a new reciprocity relation whose form is similar to the one of Eq. (15.2), that is

$$\left(\frac{s}{2i\pi}\right)^2 \int_{\kappa=-i\infty}^{i\infty} d\kappa \int_{\sigma=-i\infty}^{i\infty} \tilde{E}_x^s(\kappa, \sigma, 0, s)\partial \tilde{J}^B(-\kappa, -\sigma, s) d\sigma$$

$$= -\left(\frac{s}{2i\pi}\right)^2 \int_{\kappa=-i\infty}^{i\infty} d\kappa \int_{\sigma=-i\infty}^{i\infty} \tilde{E}_x^i(\kappa, \sigma, 0, s)\partial \tilde{J}^B(-\kappa, -\sigma, s) d\sigma$$

$$+ w\hat{\zeta}(s)\left(\frac{s}{2i\pi}\right)^2 \int_{\kappa=-i\infty}^{i\infty} \exp(s\kappa x_\zeta) d\kappa$$

$$\times \int_{\sigma=-i\infty}^{i\infty} \partial \tilde{J}^s(x_\zeta, \sigma, s)\partial \tilde{J}^B(-\kappa, -\sigma, s) d\sigma \qquad (16.2)$$

Noting, again, that the first term on the right-hand side of Eq. (16.2) has been handled in section 14.4 for the plane-wave and delta-gap excitations, we next proceed by transforming the second interaction term.

Time-Domain Electromagnetic Reciprocity in Antenna Modeling, First Edition. Martin Štumpf.
© 2020 by The Institute of Electrical and Electronics Engineers, Inc. Published 2020 by John Wiley & Sons, Inc.

16.1   MODIFICATION OF THE IMPEDITIVITY MATRIX

Assuming now that the lumped element is connected in between two discretization nodes, say $x_\zeta \in [x_Q, x_{Q+1}]$ with $Q = \{0, \ldots, N\}$, the induced electric-current surface density can be linearly interpolated in between the nodes. For the transform-domain current density that appears in the reciprocity relation (16.2), we hence get

$$\partial \tilde{J}^s(x_\zeta, \sigma, s) \simeq \sum_{k=1}^{M} \left[\Lambda^{[Q]}(x_\zeta) j_k^{[Q]} + \Lambda^{[Q+1]}(x_\zeta) j_k^{[Q+1]} \right] \tilde{\Pi}(\sigma) \hat{\Lambda}_k(s) \qquad (16.3)$$

where the interpolation weights follow, again, directly from Eq. (2.10), and, in virtue of the end conditions (14.14), we set $j_k^{[0]} = j_k^{[N+1]} = 0$. Next, $\hat{\Lambda}_k(s)$ follows from Eq. (2.11), and, finally, $\tilde{\Pi}(\sigma)$ denotes the slowness-domain counterpart of $\Pi(y)$ defined by Eq. (14.16). Substitution of Eqs. (16.3) and (14.20) in the interaction quantity yields a new impeditivity matrix, say $\mathcal{L}$, whose elements can be easily evaluated in closed form via Cauchy's formula [30, section 2.41]. By proceeding in this way, we obtain (cf. Eq. (5.4))

$$\hat{\mathcal{L}}^{[S,n]}(s) = \frac{c_0 w \hat{\zeta}(s)/s^2}{c_0 \Delta t} \left[\Lambda^{[Q]}(x_\zeta) \delta_{n,Q} + \Lambda^{[Q+1]}(x_\zeta) \delta_{n,Q+1} \right]$$
$$\times \left[H(x_\zeta + \Delta/2 - x_S) - H(x_\zeta - \Delta/2 - x_S) \right] \qquad (16.4)$$

for all $S = \{1, \ldots, N\}$ and $n = \{1, \ldots, N\}$, where $\delta_{m,n}$ is the Kronecker delta. The TD counterpart of Eq. (16.4) can be written in the following form

$$\mathcal{L}^{[S,n]}(t) = \mathcal{F}(t) \left[\Lambda^{[Q]}(x_\zeta) \delta_{n,Q} + \Lambda^{[Q+1]}(x_\zeta) \delta_{n,Q+1} \right]$$
$$\times \left[H(x_\zeta + \Delta/2 - x_S) - H(x_\zeta - \Delta/2 - x_S) \right] \qquad (16.5)$$

in which $\mathcal{F}(t)$ is the TD counterpart of $w \hat{\zeta}(s)/s^2 \, \Delta t$. Upon inspection with $F(t)$ as defined in section 5.1, it is clear that $\mathcal{F}(t) = w F(t)$. Consequently, we get

$$\mathcal{F}(t) = w R \, (c_0 t / c_0 \Delta t) \, H(t) \qquad (16.6)$$

for a resistor of resistance R and

$$\mathcal{F}(t) = (w \Delta t / 2C)(c_0 t / c_0 \Delta t)^2 \, H(t) \qquad (16.7)$$

for a capacitor of capacitance C, and finally,

$$\mathcal{F}(t) = (w L / \Delta t) \, H(t) \qquad (16.8)$$

for an inductor of inductance L. Following the methodology applied in section 15.1, the impact of a lumped element on the electric-current surface density is incorporated by replacing the impeditivity matrix $\underline{Z}$ in Eq. (14.22) with $\underline{Z} - \underline{L}$. Finally, recall that the elements of the impeditivity array $\underline{Z}$ can be evaluated with the help of the results summarized in appendix G.

CHAPTER 17

INCLUDING A PEC GROUND PLANE

In section 11.1, it has been demonstrated that the electric-current distribution along a gap-excited wire above a PEC ground can be approximately described with the aid of transmission-line theory. Whenever the corresponding "low-frequency assumptions" are no longer applicable, one has to resort to an alternative solution strategy. Therefore, developing a numerical scheme that is capable of calculating the response of a narrow strip antenna over a PEC ground is the main objective of the present chapter. Namely, relying heavily on the results presented in chapter 14, it is shown that the CdH-MoM formulation can be readily generalized to account for the effect of the ground plane. Illustrative numerical examples concerning both solutions based on the CdH-MoM and transmission-line theory are discussed.

17.1 PROBLEM DESCRIPTION

We shall analyze the response of a narrow PEC strip antenna that is located above a perfect ground plane (see Figure 17.1). Again, the planar antenna under consideration occupies a bounded domain in $\{-\ell/2 < x < \ell/2, -w/2 < y < w/2, z = 0\}$, where $\ell > 0$ denotes its length and $w > 0$ is its width. The closed surface separating the strip antenna from its exterior domain is denoted S_0, again. The (unbounded) PEC ground plane extends over $\{-\infty < x < \infty, -\infty < y < \infty, z = -z_0\}$, where $z_0 > 0$. The antenna radiates into the half-space $\mathcal{D}_+^\infty$, whose EM properties are described by (real-valued, positive and scalar) electric permittivity ϵ_0 and magnetic permeability μ_0. The planar antenna is excited by uniform EM plane wave or/and via a localized voltage source in the narrow gap located at $\{x_\delta - \delta/2 < x < x_\delta + \delta/2, -w/2 < y < w/2, z = 0\}$ with $\delta > 0$ denoting its (vanishing) width. The scattered wave fields, $\{\hat{\boldsymbol{E}}^s, \hat{\boldsymbol{H}}^s\}$, are defined as secondary fields that are radiated from induced currents along the conducting strip (see Eq. (11.12)).

Time-Domain Electromagnetic Reciprocity in Antenna Modeling, First Edition. Martin Štumpf.
© 2020 by The Institute of Electrical and Electronics Engineers, Inc. Published 2020 by John Wiley & Sons, Inc.

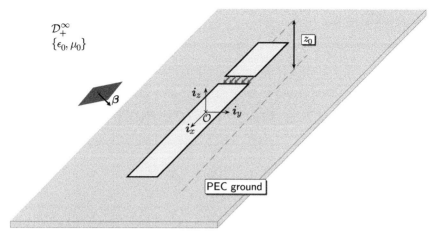

FIGURE 17.1. A planar-strip antenna over the ground plane.

17.2 PROBLEM FORMULATION

Again, the problem will be tackled with the aid of the EM reciprocity theorem of the time-convolution type [4, section 1.4.1]. To that end, the testing EM state is chosen to be associated with the testing electric-current surface density distributed along the strip in presence of the perfect ground. Hence, applying the Lorentz reciprocity theorem to the half-space $\mathcal{D}_+^\infty$ excluding $\mathcal{S}_0$ and to the scattered-field (s) and testing-field (B) states (see Table 17.1), we end up with the reciprocity relation (14.8), again, where we used the explicit-type boundary conditions applying over the ground plane, that is

$$\lim_{z\downarrow-z_0} \{\hat{E}_x^{s,B}, \hat{E}_y^{s,B}\}(x,y,z,s) = \{0,0\} \tag{17.1}$$

for all $x \in \mathbb{R}$ and $y \in \mathbb{R}$. The slowness-domain counterpart of the starting reciprocity relation has then the form as Eq. (14.10), that is

$$\left(\frac{s}{2i\pi}\right)^2 \int_{\kappa=-i\infty}^{i\infty} d\kappa \int_{\sigma=-i\infty}^{i\infty} \tilde{E}_x^B(\kappa,\sigma,0,s)\partial\tilde{J}^s(-\kappa,-\sigma,s)d\sigma$$

$$= \left(\frac{s}{2i\pi}\right)^2 \int_{\kappa=-i\infty}^{i\infty} d\kappa \int_{\sigma=-i\infty}^{i\infty} \tilde{E}_x^s(\kappa,\sigma,0,s)\partial\tilde{J}^B(-\kappa,-\sigma,s)d\sigma \quad \text{(14.10 revisited)}$$

in which the testing field $\tilde{E}_x^B$ is modified such that Eq. (17.1) applies. Following this reasoning, we arrive at (cf. Eq. (14.11))

$$\tilde{E}_x^B(\kappa,\sigma,z,s) = -(s/\epsilon_0)\Omega_0^2(\kappa)\tilde{K}(\kappa,\sigma,z,s)\partial\tilde{J}^B(\kappa,\sigma,s) \tag{17.2}$$

TABLE 17.1. Application of the Reciprocity Theorem

Domain $\mathcal{D}_+^\infty \setminus \mathcal{S}_0$		
Time-Convolution	State (s)	State (B)
Source	0	0
Field	$\{\hat{\boldsymbol{E}}^s, \hat{\boldsymbol{H}}^s\}$	$\{\hat{\boldsymbol{E}}^B, \hat{\boldsymbol{H}}^B\}$
Material	$\{\epsilon_0, \mu_0\}$	$\{\epsilon_0, \mu_0\}$

in which the (transform-domain) Green's function accounts for the reflections from the perfect ground plane, that is

$$\tilde{K}(\kappa, \sigma, z, s) = \tilde{G}(\kappa, \sigma, z, s) - \tilde{G}(\kappa, \sigma, z + 2z_0, s) \tag{17.3}$$

where we use Eq. (14.12) and recall that $\Omega_0^2(\kappa) = 1/c_0^2 - \kappa^2$. The transform-domain reciprocity relation (14.10) (see above Eq. (17.2)) with the modified testing field (see Eqs. (17.2) with (17.3)) is the point of departure for the Cagniard-DeHoop Method of Moments (CdH-MoM) solution given in the next section.

17.3 PROBLEM SOLUTION

As in section 14.3, the spatial solution domain extends over the surface of the strip antenna. Adopting the strategy applied in section 14.3, the surface is divided into rectangular elements over which the electric-current distribution is approximated by a set of piecewise linear functions (see Figure 14.2). Thanks to the problem linearity, an inspection of Eqs. (14.11) and (14.12) with Eqs. (17.2) and (17.3) reveals that the impact of reflections from the ground plane can be captured in a separate "ground-plane impeditivity" array, say $\mathcal{K}$, whose elements are closely specified in appendix J. The electric-current surface density induced on the strip then follows from Eq. (15.8), that is

$$\boldsymbol{J}_m = \overline{\boldsymbol{\mathcal{Z}}}_1^{-1}$$

$$\cdot \left[\boldsymbol{V}_m - \sum_{k=1}^{m-1} (\overline{\boldsymbol{\mathcal{Z}}}_{m-k+1} - 2\overline{\boldsymbol{\mathcal{Z}}}_{m-k} + \overline{\boldsymbol{\mathcal{Z}}}_{m-k-1}) \cdot \boldsymbol{J}_k \right] \tag{15.8 revisited}$$

for all $m = \{1, \ldots, M\}$, where we use

$$\overline{\boldsymbol{\mathcal{Z}}} = \boldsymbol{\mathcal{Z}} - \boldsymbol{\mathcal{K}} \tag{17.4}$$

where $\boldsymbol{\mathcal{Z}}$ describes the impeditivity matrix corresponding to a PEC strip on the homogeneous, isotropic, and loss-free background (see chapter 14 and appendix G).

17.4 ANTENNA EXCITATION

In order to properly incorporate the effect of ground plane in the reciprocity-based computational model, the slowness-domain boundary condition (14.25) is replaced with

$$\tilde{E}_x^{\rm s}(\kappa, \sigma, 0, s) = -\tilde{E}_x^{\rm e}(\kappa, \sigma, 0, s) \tag{17.5}$$

where $\tilde{E}_x^{\rm e}$ denotes the x-component of the (transform-domain) excitation electric-field strength. In accordance with Eq. (11.12), the excitation field is defined as a sum of the incident EM field and the field reflected from the PEC ground. If the strip antenna under consideration is activated via the localized delta-gap source, the elements of the excitation array V_m follow directly from Eq. (14.30). The excitation via a uniform EM plane wave calls for the distribution of the tangential excitation electric field along the strip. With the explicit-type boundary condition (17.1) in mind, the desired excitation-field distribution reads (cf. Eq. (14.23))

$$\hat{E}_x^{\rm e}(x, y, 0, s) = \hat{e}^{\rm i}(s) \sin(\phi) \exp\{-s[\kappa_0(x + \ell/2) + \sigma_0 y]\}$$
$$\times \{1 - \exp[-2sz_0 \cos(\theta)/c_0]\} \tag{17.6}$$

and its slowness-domain counterpart follows using Eq. (14.24) as

$$\tilde{E}_x^{\rm e}(\kappa, \sigma, 0, s) = \hat{e}^{\rm i}(s) \exp(-s\kappa_0 \ell/2)\{1 - \exp[-2sz_0 \cos(\theta)/c_0]\}$$
$$\times \ell w \sin(\phi) i_0[s(\kappa - \kappa_0)\ell/2] i_0[s(\sigma - \sigma_0)w/2] \tag{17.7}$$

Consequently, combining Eq. (17.5) with (17.7) and substituting the result in the right-hand side of the reciprocity relation (14.10), we obtain slowness-domain integral expressions that can be carried out using Cauchy's formula [30, section 2.41]. Transforming the result of integration to the TD, we finally end up with the expression similar to Eq. (14.26), that is

$$V^{[S]}(t) = -[\sin(\phi)/\kappa_0]\{e^{\rm i}(t) - e^{\rm i}[t - 2z_0 \cos(\theta)/c_0]\}$$
$$*_t \{{\rm H}[t - \kappa_0(x_S + \ell/2 - \Delta/2)] - {\rm H}[t - \kappa_0(x_S + \ell/2 + \Delta/2)]\} \tag{17.8}$$

for all $S = \{1, \ldots, N\}$. If the incident EM plane wave propagates in the negative z-direction, then $\theta = 0$ and we get the limit (cf. Eq. (14.27))

$$V^{[S]}(t) = -[e^{\rm i}(t) - e^{\rm i}(t - 2z_0/c_0)]\Delta \sin(\phi) \tag{17.9}$$

for all $S = \{1, \ldots, N\}$, again.

ILLUSTRATIVE NUMERICAL EXAMPLE

> • Make use of the CdH-MoM to calculate the electric-current response of a gap-excited narrow strip over the PEC ground. Subsequently, employ the concept of equivalent radius [17] and compare the response with the corresponding result predicted by transmission-line theory.

Solution: We shall analyze a narrow PEC strip of width $w = 1.0$ mm and length $\ell = 100\,w$ that is located at $z_0 = 20\,w$ above the perfect ground (see Figure 17.2). The strip is at its center at $x = x_\delta = 0$ excited via a narrow gap, where the voltage pulse $V^{\mathrm{T}}(t)$ is applied. The latter has the shape of a bipolar triangle, which is described by (cf. Eq. (9.3))

$$
V^{\mathrm{T}}(t) = \frac{2V_{\mathrm{m}}}{t_{\mathrm{w}}} \left[t\,\mathrm{H}(t) - 2\left(t - \frac{t_{\mathrm{w}}}{2}\right)\mathrm{H}\left(t - \frac{t_{\mathrm{w}}}{2}\right) \right.
$$
$$
\left. +2\left(t - \frac{3t_{\mathrm{w}}}{2}\right)\mathrm{H}\left(t - \frac{3t_{\mathrm{w}}}{2}\right) - (t - 2t_{\mathrm{w}})\mathrm{H}(t - 2t_{\mathrm{w}}) \right] \qquad (17.10)
$$

with $c_0 t_{\mathrm{w}} = 1.0\,\ell$ and $V_{\mathrm{m}} = 1.0$ V. The corresponding pulse shape is then similar to the one shown in Figure 7.2b, but with the unit voltage amplitude. EM properties of the half-space above the ground are described by $\{\epsilon_0, \mu_0\}$, thus implying the wave speed $c_0 = (\epsilon_0\mu_0)^{-1/2} > 0$.

In the first step, we recall the conclusions drawn in section 11.1 and analyze the problem approximately with the aid of transmission-line theory. To that end, we employ Eq. (14.51) to find the equivalent radius of the transmission line, that is, $a = w \exp(-3/2)$. The electric-current and voltage distribution along such a line is then (approximately) governed by transmission-line equations (cf. Eqs. (11.10) and (11.11))

$$
\partial_x \hat{V}(x, s) + \hat{\zeta}(s)\hat{I}(x, s) = 0 \qquad (17.11)
$$

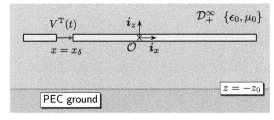

FIGURE 17.2. A gap-excited strip above the perfect ground.

$$\partial_x \hat{I}(x, s) + \hat{\eta}(s)\hat{V}(x, s) = 0 \tag{17.12}$$

where the longitudinal impedance $\hat{\zeta}(s)$ and the transverse admittance $\hat{\eta}(s)$ are given by Eqs. (11.6) and (11.7). The voltage-gap excitation located at $x = x_\delta$ is incorporated via the excitation condition

$$\lim_{\delta \downarrow 0} \hat{V}(x_\delta + \delta/2, s) - \lim_{\delta \downarrow 0} \hat{V}(x_\delta - \delta/2, s) = \hat{V}^{\mathrm{T}}(s) \tag{17.13}$$

while

$$\lim_{\delta \downarrow 0} \hat{I}(x_\delta + \delta/2, s) - \lim_{\delta \downarrow 0} \hat{I}(x_\delta - \delta/2, s) = 0 \tag{17.14}$$

across the gap. Furthermore, since the strip under consideration is open-circuited at its ends, the transmission-line equations are further supplemented with the corresponding end conditions (cf. Eq. (2.9))

$$\hat{I}(\pm \ell/2, s) = 0 \tag{17.15}$$

The electric-current distribution along the open-ended line is sought in the following form

$$\hat{I}(x, s) = \begin{cases} A \exp[-s(x - x_\delta)/c_0] + B \exp[-s(\ell/2 - x_\delta)/c_0], \\ C \exp[s(x - x_\delta)/c_0] + D \exp[-s(\ell/2 + x_\delta)/c_0] \end{cases} \tag{17.16}$$

for $\{x_\delta \leq x \leq \ell/2\}$ and $\{-\ell/2 \leq x \leq x_\delta\}$, respectively, where A, B, C, and D are, yet unknown, coefficients. It is straightforward to show that the boundary conditions Eqs. (17.13)–(17.15) are met, if

$$A = \frac{Y^c \hat{V}^{\mathrm{T}}(s)}{2} \frac{1 - \exp[-2s(\ell/2 + x_\delta)/c_0]}{1 - \exp(-2s\ell/c_0)} \tag{17.17}$$

$$B = -\frac{Y^c \hat{V}^{\mathrm{T}}(s)}{2} \frac{\exp[-s(\ell/2 - x_\delta)/c_0] - \exp[-s(3\ell/2 + x_\delta)/c_0]}{1 - \exp(-2s\ell/c_0)} \tag{17.18}$$

$$C = \frac{Y^c \hat{V}^{\mathrm{T}}(s)}{2} \frac{1 - \exp[-2s(\ell/2 - x_\delta)/c_0]}{1 - \exp(-2s\ell/c_0)} \tag{17.19}$$

$$D = -\frac{Y^c \hat{V}^{\mathrm{T}}(s)}{2} \frac{\exp[-s(\ell/2 + x_\delta)/c_0] - \exp[-s(3\ell/2 - x_\delta)/c_0]}{1 - \exp(-2s\ell/c_0)} \tag{17.20}$$

where $Y^c = 2\pi\eta_0/\ln(2z_0/a)$ is the characteristic admittance of the transmission line (see Eq. (11.9)). Expanding next the coefficients in a (convergent) geometric series via

$$\frac{1}{1 - \exp(-2s\ell/c_0)} = \sum_{n=0}^{\infty} \exp(-2sn\ell/c_0) \tag{17.21}$$

the electric-current distribution along the line can be expressed as

$$\hat{I}(x, s) = \sum_{n=0}^{\infty} \hat{I}^{[n]}(x, s) \tag{17.22}$$

where the expansion constituents are given by

$$\hat{I}^{[n]}(x, s) = \hat{I}^{\mathrm{T};[n]}(s)\{\exp[-s(x - x_\delta)/c_0] - \exp[-s(\ell - x - x_\delta)/c_0]$$
$$+ \exp[-s(2\ell - x + x_\delta)/c_0] - \exp[-s(\ell + x + x_\delta)/c_0]\} \tag{17.23}$$

for $\{x_\delta \leq x \leq \ell/2\}$ and

$$\hat{I}^{[n]}(x, s) = \hat{I}^{\mathrm{T};[n]}(s)\{\exp[-s(x_\delta - x)/c_0] - \exp[-s(\ell + x_\delta + x)/c_0]$$
$$+ \exp[-s(2\ell + x - x_\delta)/c_0] - \exp[-s(\ell - x - x_\delta)/c_0]\} \tag{17.24}$$

for $\{-\ell/2 \leq x \leq x_\delta\}$, where we defined

$$\hat{I}^{\mathrm{T};[n]}(s) = \frac{Y^c \hat{V}^{\mathrm{T}}(s)}{2} \exp(-2sn\ell/c_0) \tag{17.25}$$

Subsequently, the current distribution is transformed back to the TD, which yields

$$I(x, t) = \sum_{n=0}^{\infty} I^{[n]}(x, t) \tag{17.26}$$

with

$$I^{[n]}(x, t) = I^{\mathrm{T};[n]}[t - (x - x_\delta)/c_0] - I^{\mathrm{T};[n]}[t - (\ell - x - x_\delta)/c_0]$$
$$+ I^{\mathrm{T};[n]}[t - (2\ell - x + x_\delta)/c_0] - I^{\mathrm{T};[n]}[t - (\ell + x + x_\delta)/c_0] \tag{17.27}$$

for $\{x_\delta \leq x \leq \ell/2\}$ and

$$I^{[n]}(x, t) = I^{\mathrm{T};[n]}[t - (x_\delta - x)/c_0] - I^{\mathrm{T};[n]}[t - (\ell + x + x_\delta)/c_0]$$
$$+ I^{\mathrm{T};[n]}[t - (2\ell + x - x_\delta)/c_0] - I^{\mathrm{T};[n]}[t - (\ell - x - x_\delta)/c_0] \tag{17.28}$$

for $\{-\ell/2 \leq x \leq x_\delta\}$. Finally, the TD counterpart of Eq. (17.25) follows from

$$I^{\mathrm{T};[n]}(t) = Y^c V^{\mathrm{T}}(t - 2n\ell/c_0)/2 \tag{17.29}$$

Apparently, since $V^{\mathrm{T}}(t) = 0$ for $t \leq 0$, the number of terms in Eq. (17.26) is finite in any bounded time window of observation.

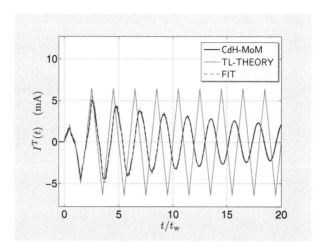

FIGURE 17.3. Electric-current self-responses as calculated using the CdH-MoM, transmission-line theory (TL-THEORY) and FIT.

The electric-current (self)-responses of the analyzed antenna observed at $x = 0$ are shown in Figure 17.3. As can be seen, the responses calculated with the aid of the CdH-MoM (see sections 17.2 and 17.3)) and using Eq. (17.26) with $I^T(t) = I(0, t)$ do differ significantly except for the early-time part of the response. The discrepancies between the calculated responses can largely be attributed to a relatively large distance of the strip from the ground plane, which is $z_0 = c_0 t_w/5$. Recall that the line-to-ground distance should be sufficiently small with respect to the spatial support of the excitation pulse for transmission-line theory to apply (see Eq. (11.3)). For the sake of validation, the electric-current response has finally been calculated with the help of the Finite Integration Technique (FIT) as implemented in CST Microwave Studio. As can be observed, the resulting pulse shape agrees with the one obtained using the CdH-MoM.

APPENDIX A

A GREEN'S FUNCTION REPRESENTATION IN AN UNBOUNDED, HOMOGENEOUS, AND ISOTROPIC MEDIUM

The Green's function of the scalar Helmholtz equation can be represented via an inverse spatial Fourier integral in the three-dimensional angular wave-vector domain [3, section 26.2]. Assuming a spatial Fourier transform along the z-direction only, the inverse Fourier transform of the Green's function may be written in the following form

$$\hat{G}(r, z, s) = \frac{s}{2i\pi} \int_{p=-i\infty}^{i\infty} \tilde{G}(r, p, s) \exp(-spz) dp \qquad (A.1)$$

where $r = (x^2 + y^2)^{1/2} > 0$, p is the wave-slowness parameter, and s denotes the real-valued (scaling) Laplace-transform parameter. Since the slowness representation (A.1) entails $\partial_z \to -sp$, the transform-domain Green's function satisfies the two-dimensional, azimuth-independent Helmholtz equation

$$r^{-1}\partial_r[r\partial_r\tilde{G}(r, p, s)] - s^2\gamma_0^2(p)\tilde{G}(r, p, s) = -\delta(r)/2\pi r \qquad (A.2)$$

with $\gamma_0^2(p) = 1/c_0^2 - p^2$, whose solution can be described by a modified Bessel function of the second kind, that is,

$$\tilde{G}(r, p, s) = K_0[s\gamma_0(p)r]/2\pi \qquad (A.3)$$

Time-Domain Electromagnetic Reciprocity in Antenna Modeling, First Edition. Martin Štumpf.
© 2020 by The Institute of Electrical and Electronics Engineers, Inc. Published 2020 by John Wiley & Sons, Inc.

In virtue of causality, the solution is bounded as $r \to \infty$ with $\mathrm{Re}[\gamma_0(p)] > 0$, the latter being consistent with the principal branch of the modified Bessel function. Combining next Eq. (A.1) with Eq. (A.3), we end up with

$$\frac{\exp(-sR/c_0)}{4\pi R} = \frac{s}{4\pi^2 \mathrm{i}} \int_{p=-\mathrm{i}\infty}^{\mathrm{i}\infty} K_0[s\gamma_0(p)r] \exp(-spz)\mathrm{d}p \qquad (A.4)$$

where $R = (x^2 + y^2 + z^2)^{1/2} > 0$, which is closely related to the "modified Sommerfeld's integral" whose importance in EM theory has been recognized by Schelkunoff (see [48] and [49, p. 414]).

Schelkunoff's identity (A.4) can also be established with the aid of the CdH technique [8]. To that end, we make use of the integral representation of the modified Bessel function [25, eq. (9.6.23)] and rewrite the right-hand side of (A.4) as

$$\frac{s}{4\pi^2 \mathrm{i}} \int_{u=1}^{\infty} \frac{\mathrm{d}u}{(u^2-1)^{1/2}} \int_{p=-\mathrm{i}\infty}^{\mathrm{i}\infty} \exp\{-s[pz + \gamma_0(p)ur]\}\mathrm{d}p \qquad (A.5)$$

In the next step, the integration contour along the imaginary p axis is replaced by the CdH path that is defined as

$$\tau = pz + \gamma_0(p)ur \qquad (A.6)$$

where $\{\tau \in \mathbb{R}; 0 < T(u) \le \tau < \infty\}$, with $T(u) = D(u)/c_0$ and $D(u) = (z^2 + u^2 r^2)^{1/2} > 0$. Combining further the contributions of integration in the upper and lower halves of the complex p-plane and introducing the τ-parameter as a new variable of integration, we find that Eq. (A.5) can be written as

$$\frac{sr}{2\pi^2} \int_{u=1}^{\infty} \frac{u\,\mathrm{d}u}{(u^2-1)^{1/2}} \int_{\tau=T(u)}^{\infty} \frac{\exp(-s\tau)}{D^2(u)} \frac{\tau\,\mathrm{d}\tau}{[\tau^2 - T^2(u)]^{1/2}} \qquad (A.7)$$

where we have used the values of the Jacobian $\partial p/\partial \tau = \mathrm{i}\gamma_0[p(\tau)]/[\tau^2 - T^2(v)]^{1/2}$ along the (hyperbolic) CdH path in $\mathrm{Im}(p) > 0$. After changing the order of the integrations, we get

$$\frac{sc_0}{2\pi^2} \int_{\tau=R/c_0}^{\infty} \tau \exp(-s\tau)\mathrm{d}\tau$$

$$\times \int_{u=1}^{Q(\tau)} \frac{1}{D^2(u)} \frac{u\,\mathrm{d}u}{[Q^2(\tau) - u^2]^{1/2}(u^2-1)^{1/2}} \qquad (A.8)$$

where $Q(\tau) = (c_0^2 \tau^2 - z^2)^{1/2}/r > 0$. The inverse square-root singularities in the lower and upper limits of the integration with respect to u can be handled via stretching the variable of integration. Accordingly, we substitute

$$u^2 = \cos^2(\psi) + Q^2(\tau)\sin^2(\psi) \qquad (A.9)$$

for $\{0 \leq \psi \leq \pi/2\}$ and rewrite Eq. (A.8) as

$$\frac{sc_0}{2\pi^2}\int_{\tau=R/c_0}^{\infty}\tau\exp(-s\tau)\mathrm{d}\tau\int_{\psi=0}^{\pi/2}\frac{\mathrm{d}\psi}{R^2\cos^2(\psi)+c_0^2\tau^2\sin^2(\psi)} \qquad \text{(A.10)}$$

The inner integral with respect to ψ is a standard integral that can be evaluated analytically, and we finally end up with

$$\frac{s}{4\pi R}\int_{\tau=R/c_0}^{\infty}\exp(-s\tau)\mathrm{d}\tau = \frac{\exp(-sR/c_0)}{4\pi R} \qquad \text{(A.11)}$$

which agrees with the left-hand side of (A.4), thus proving the identity.

APPENDIX B

TIME-DOMAIN RESPONSE OF AN INFINITE CYLINDRICAL ANTENNA

A limited number of problems in EM theory can be solved analytically. An example from this category is the description of the electric-current distribution along a gap-excited, infinitely-long PEC cylindrical antenna (see [50–52]). In this appendix, it is shown how this problem can be handled with the help of the CdH technique [8].

B.1 TRANSFORM-DOMAIN SOLUTION

We shall analyze the cylindrical antenna shown in Figure B.1. To this end, the total field in the problem configuration is first written as the superposition of the scattered (s) EM field and the incident (i) EM field localized in the excitation gap. Exterior to the domain occupied by the antenna, the scattered (s) field is governed by a ϕ-independent set of EM equations in the complex FD

$$\partial_z \hat{H}_\phi^s = -s\epsilon_0 \hat{E}_r^s \tag{B.1}$$

$$r^{-1}\partial_r(r\hat{H}_\phi^s) = s\epsilon_0 \hat{E}_z^s \tag{B.2}$$

$$\partial_z \hat{E}_r^s - \partial_r \hat{E}_z^s = -s\mu_0 \hat{H}_\phi^s \tag{B.3}$$

where $\hat{E}_z^s$ and $\hat{E}_r^s$ are the axial and radial components of the electric-field strength, respectively, and $\hat{H}_\phi^s$ denotes the azimuthal component of the corresponding magnetic-field strength. The EM field equations are further supplemented with the excitation condition applying in the narrow gap

$$\lim_{r \downarrow a} \hat{E}_z^s(r, z, s) = -\hat{V}^T(s)\delta(z) \tag{B.4}$$

Time-Domain Electromagnetic Reciprocity in Antenna Modeling, First Edition. Martin Štumpf.
© 2020 by The Institute of Electrical and Electronics Engineers, Inc. Published 2020 by John Wiley & Sons, Inc.

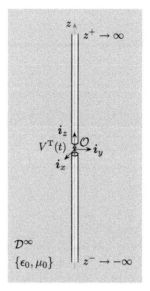

FIGURE B.1. A straight, infinite-wire antenna excited by a voltage gap source.

for all $z \in \mathbb{R}$. Referring now to identity (A.4), it is postulated that the axial component of the electric-field strength can be represented via the slowness-domain integral

$$\hat{E}_z^{\mathrm{s}}(r, z, s) = \frac{s\hat{V}^{\mathrm{T}}(s)}{2\pi\mathrm{i}} \int_{p=-\mathrm{i}\infty}^{\mathrm{i}\infty} \tilde{A}(p, s)\mathrm{K}_0[s\gamma_0(p)r]\exp(-spz)\mathrm{d}p \tag{B.5}$$

Using the slowness representation of the Dirac-delta distribution in (B.4), combination of Eqs. (B.4) and (B.5) leads to

$$\tilde{A}(p, s) = -1/\mathrm{K}_0[s\gamma_0(p)a] \tag{B.6}$$

The latter implies that the transform-domain solution is given by $\tilde{E}_z^{\mathrm{s}}(r, p, s) = -\hat{V}^{\mathrm{T}}(s)\mathrm{K}_0[s\gamma_0(p)r]/\mathrm{K}_0[s\gamma_0(p)a]$, which can be further used to express the transform-domain electric-current distribution along the antenna, that is

$$\tilde{I}(p, s) = 2\pi a \lim_{r\downarrow a} \tilde{H}_\phi(r, p, s) = 2\pi a \frac{s\epsilon_0}{s^2\gamma_0^2(p)} \lim_{r\downarrow a} \partial_r \tilde{E}_z(r, p, s) \tag{B.7}$$

where we have employed the combination of Eqs. (B.1) and (B.3) in the transform domain. In this way, we finally obtain

$$\tilde{I}(p, s) = 2\pi a \hat{V}^{\mathrm{T}}(s) \frac{\epsilon_0}{\gamma_0(p)} \frac{\mathrm{K}_1[s\gamma_0(p)a]}{\mathrm{K}_0[s\gamma_0(p)a]} \tag{B.8}$$

The transformation of (B.8) to the TD is the subject of the following section.

B.2 TIME-DOMAIN SOLUTION

The wave-slowness representation of the induced electric current along the infinite wire can be found with the help of (B.8) as

$$\hat{I}(z,s) = -is\epsilon_0 a \hat{V}^{\mathrm{T}}(s) \int_{p=-i\infty}^{i\infty} \frac{K_1[s\gamma_0(p)a]}{K_0[s\gamma_0(p)a]} \exp(-spz) \frac{dp}{\gamma_0(p)} \tag{B.9}$$

In order to transform (B.9) to the TD, we deform the integration contour along the imaginary p axis into a new path, along which the radial slowness coefficient $\gamma_0(p)$ is purely imaginary, that is

$$s\gamma_0(p)a = \pm i\zeta \tag{B.10}$$

where $\{\zeta \in \mathbb{R}; \zeta \geq 0\}$. The deformation is permissible in virtue of Jordan's lemma and Cauchy's theorem. Solving (B.10) for p, we get a representation of the loop around the branch cuts extending along $\{\mathrm{Im}(p) = 0, 1/c_0 \leq |\mathrm{Re}(p)| < \infty\}$, that is

$$p(\zeta) = \pm(sc_0)^{-1}(s^2 + c_0^2\zeta^2/a^2)^{1/2} \tag{B.11}$$

in which $+$ applies to $z > 0$ and $-$ to $z < 0$. The Jacobian of the mapping follows

$$\frac{\partial p}{\partial \zeta} = \pm \frac{c_0}{s} \frac{\zeta/a^2}{(s^2 + c_0^2\zeta^2/a^2)^{1/2}} \tag{B.12}$$

Combining the contributions of integration just above and below the branch cut, we arrive at

$$\hat{I}(z,s) = 2\eta_0 s \hat{V}^{\mathrm{T}}(s) \int_{\zeta=0}^{\infty} \frac{\exp[-(s^2 + c_0^2\zeta^2/a^2)^{1/2}|z|/c_0]}{(s^2 + c_0^2\zeta^2/a^2)^{1/2}}$$
$$\mathrm{Re}\{K_1(-i\zeta)/K_0(-i\zeta)\}d\zeta \tag{B.13}$$

where (the real part of) the ratio of the modified Bessel functions can be further expressed in terms of standard Bessel functions, that is

$$\hat{I}(z,s) = \frac{4}{\pi}\eta_0 s \hat{V}^{\mathrm{T}}(s) \int_{\zeta=0}^{\infty} \frac{\exp[-(s^2 + c_0^2\zeta^2/a^2)^{1/2}|z|/c_0]}{(s^2 + c_0^2\zeta^2/a^2)^{1/2}}$$
$$\frac{1}{Y_0^2(\zeta) + J_0^2(\zeta)} \frac{d\zeta}{\zeta} \tag{B.14}$$

Finally, the TD counterpart of the electric-current distribution follows upon applying [25, eq. (9.6.23)], that is

$$I(z,t) = \frac{4}{\pi}\eta_0 \partial_t V^{\mathrm{T}}(t) *_t \int_{\zeta=0}^{\infty} \frac{J_0(\zeta\tau)}{Y_0^2(\zeta) + J_0^2(\zeta)} \frac{d\zeta}{\zeta} H(c_0 t - z) \tag{B.15}$$

where $\tau = (c_0^2 t^2 - z^2)^{1/2}/a$. Equation (B.15) is clearly equivalent to the results derived before through different approaches (see [52, eq. (17)] and [51, eq. (9)]).

Alternatively, we may use

$$\frac{K_1(-i\zeta)}{K_0(-i\zeta)} = -i\frac{d}{d\zeta}\ln[K_0(-i\zeta)] \tag{B.16}$$

in Eq. (B.13), which upon applying integration by parts yields another TD expression

$$I(z,t) = -2\eta_0 \partial_t V^{\mathrm{T}}(t) *_t \tau \int_{\zeta=0}^{\infty} J_1(\zeta\tau)\tan^{-1}[J_0(\zeta)/Y_0(\zeta)]d\zeta \tag{B.17}$$

for $t > |z|/c_0$. If τ is large, the major contribution to the integration comes from $\zeta \downarrow 0$, which motivates to use $J_0(\zeta) = 1 + O(\zeta^2)$ and $Y_0(\zeta) = (2/\pi)[\ln(\zeta/2) + \gamma] + O(\zeta^2)$ in Eq. (B.17). This way leads to the starting relation for deriving a useful asymptotic expansion (see [50, eq. (5)]). Yet alternatively, one may express the current distribution in terms of (nonoscillating) modified Bessel functions, which is convenient for numerical purposes. To this end, we first express $J_0(\zeta\tau)$ in terms of K_0-functions such that we get the difference of two terms

$$I(z,t) = \frac{2}{\pi}\eta_0 \partial_t V^{\mathrm{T}}(t) *_t \mathrm{Im}\int_{\zeta=0}^{\infty}[K_0(-i\zeta\tau) - K_0(i\zeta\tau)]$$

$$[K_1(-i\zeta)/K_0(-i\zeta)]d\zeta \tag{B.18}$$

The large-argument behavior of the modified Bessel functions then allows deforming the integration contour in the complex ζ-plane, and we get

$$\int_{\zeta=0}^{\infty}K_0(-i\zeta\tau)\frac{K_1(-i\zeta)}{K_0(-i\zeta)}d\zeta = i\int_{\nu=0}^{\infty}K_0(\nu\tau)\frac{K_1(\nu)}{K_0(\nu)}d\nu \tag{B.19}$$

$$\int_{\zeta=0}^{\infty}K_0(i\zeta\tau)\frac{K_1(-i\zeta)}{K_0(-i\zeta)}d\zeta = i\int_{\nu=0}^{\infty}K_0(\nu\tau)\frac{K_1(\nu) + i\pi I_1(\nu)}{K_0(\nu) - i\pi I_0(\nu)}d\nu \tag{B.20}$$

Substitution of Eqs. (B.19) and (B.20) in Eq. (B.18) yields the electric-distribution in the following form

$$I(z,t) = 2\pi\eta_0\partial_t V^{\mathrm{T}}(t) *_t \int_{\nu=0}^{\infty}\frac{I_0(\nu)}{K_0(\nu)}\frac{K_0(\nu\tau)}{K_0^2(\nu) + \pi^2 I_0^2(\nu)}\frac{d\nu}{\nu} \tag{B.21}$$

which is the formula introduced in [52, eq. (18)]. The latter can be further rewritten in the form that does not involve the exponentially growing function $I_0(\nu)$, that is

$$I(z,t) = 2\pi\eta_0\partial_t V^{\mathrm{T}}(t)$$

$$*_t \int_{\nu=0}^{\infty}\frac{I_0(\nu)}{K_0(\nu)}\frac{K_0(\nu\tau)}{K_0^2(\nu)\exp(-2\nu) + \pi^2 I_0^2(\nu)}\frac{d\nu}{\nu} \tag{B.22}$$

where $I_0(\nu) = I_0(\nu)\exp(-\nu)$ and $K_0(\nu) = K_0(\nu)\exp(\nu)$ denote the scaled modified Bessel functions of the first and second kind, respectively [25, figure 9.8]. Finally note that the integration just around $\nu = 0$ can be approximated with the aid of small-argument expansions of the modified Bessel functions [25, eqs. (9.6.12) and (9.6.13)]. In this way, end up with

$$\lim_{\varepsilon\downarrow 0}\int_{\nu=0}^{\varepsilon}\frac{I_0(\nu)}{K_0(\nu)}\frac{K_0(\nu\tau)}{K_0^2(\nu)+\pi^2 I_0^2(\nu)}\frac{d\nu}{\nu} \simeq \Psi[\ln(\epsilon/2)+\gamma,\tau]+\frac{1}{2} \qquad (B.23)$$

where

$$\Psi(x,\tau) = \tan^{-1}(x/\pi)/\pi + \ln(\tau)\ln[x/(x^2+\pi^2)^{1/2}]/\pi^2. \qquad (B.24)$$

APPENDIX C

IMPEDANCE MATRIX

The elements of the TD impedance matrix that appears in Eq. (2.15) are found upon transforming the following expression to the TD

$$\hat{Z}^{[S,n]}(s) = \frac{\zeta_0/2\pi}{c_0 \Delta t \Delta^2} \left\{ \frac{c_0^2}{2i\pi} \int_{p=-i\infty}^{i\infty} \frac{\tilde{E}^{[S,n]}(p,s)}{s^4 p^2} K_0[s\gamma_0(p)a] dp \right.$$

$$\left. - \frac{1}{2i\pi} \int_{p=-i\infty}^{i\infty} \frac{\tilde{E}^{[S,n]}(p,s)}{s^4 p^4} K_0[s\gamma_0(p)a] dp \right\} \qquad \text{(C.1)}$$

for all $S = \{1, \ldots, N\}$ and $n = \{1, \ldots, N\}$, where $\zeta_0 = (\mu_0/\epsilon_0)^{1/2} > 0$ is the wave impedance and

$$\tilde{E}^{[S,n]}(p,s) = [\exp(2sp\Delta) - 4\exp(sp\Delta)$$

$$+ 6 - 4\exp(-sp\Delta) + \exp(-2sp\Delta)] \exp[sp(z_S - z_n)] \qquad \text{(C.2)}$$

Furthermore, recall that $\gamma_0(p) = (1/c_0^2 - p^2)^{1/2}$ with $\text{Re}[\gamma_0(p)] \geq 0$ as given in appendix A1. Since the integrands in Eq. (C.1) have no singularity at $p = 0$, the integration contour can be indented to the right with a semicircular arc with center at the origin and a vanishingly small radius (see Figure C.1). This leads for the integration to the same result. Consequently, the matrix elements can be composed of constituents, generic integral representations of which are analyzed in the following sections.

C.1 GENERIC INTEGRAL I^A

With reference to Eqs. (C.1) and (C.2), the first generic integral to be evaluated has the form

$$\hat{I}^A(z,s) = \frac{1}{2i\pi} \int_{p \in I_0} \frac{\exp(spz)}{s^4 p^4} K_0[s\gamma_0(p)a] dp \qquad \text{(C.3)}$$

Time-Domain Electromagnetic Reciprocity in Antenna Modeling, First Edition. Martin Štumpf.
© 2020 by The Institute of Electrical and Electronics Engineers, Inc. Published 2020 by John Wiley & Sons, Inc.

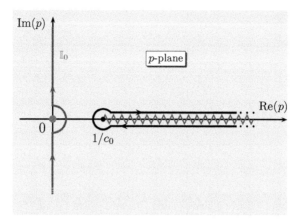

FIGURE C.1. Complex p-plane with the original integration contour $\mathbb{I}_0$ and the new CdH path for $z < 0$ encircling the branch cut.

for $z \in \mathbb{R}$ and $\{s \in \mathbb{R}; s > 0\}$, where the integration path $\mathbb{I}_0$ is depicted in Figure C.1. To find the TD counterpart of Eq. (C.3), we first note that the integrand has a fourfold pole singularity at $p = 0$. In addition, we identify the branch cuts along $\{\mathrm{Im}(p) = 0, 1/c_0 \leq |\mathrm{Re}(p)| < \infty\}$ due to the square root in $\gamma_0(p)$. Referring to the CdH technique [8], the integration contour $\mathbb{I}_0$ is next replaced by the CdH path that is defined by

$$p(\tau) = -\tau/z \pm i0 \tag{C.4}$$

for $\{\tau \in \mathbb{R}; |z|/c_0 \leq \tau < \infty\}$, which represents the loops encircling the branch cuts. It is noted that the contour deformation is permissible since the integrand is $o(1)$ as $|p| \to \infty$, thus meeting the condition for Jordan's lemma to apply [3, p. 1054]. Subsequently, combining the integrations just below and just above the branch cut, we find

$$\hat{I}^A(z, s) = \frac{|z|^3}{\pi} \int_{\tau=|z|/c_0}^{\infty} \frac{\exp(-s\tau)}{s^4} \mathrm{Im}\{\mathrm{K}_0[-isa(\tau^2/z^2 - 1/c_0^2)^{1/2}]\} \frac{d\tau}{\tau^4}$$
$$+ \left[\frac{z^3}{6} \frac{\mathrm{K}_0(sa/c_0)}{s} + \frac{azc_0}{2} \frac{\mathrm{K}_1(sa/c_0)}{s^2} \right] \mathrm{H}(z) \tag{C.5}$$

where we have included, in virtue of Cauchy's theorem, the contribution of pole at $p = 0$. In view of the chosen contour indentation (see Figure C.1), the latter is nonzero whenever $z > 0$. If the radius of the antenna is relatively small, the integrand can be considerably simplified with the help of $\mathrm{Im}[\mathrm{K}_0(-ix)] = (\pi/2)\mathrm{J}_0(x) = \pi/2 + O(x^2)$ as $x \downarrow 0$. The latter leads to

$$\hat{I}^A(z,s) \simeq \frac{|z|^3}{2} \int_{\tau=|z|/c_0}^{\infty} \frac{\exp(-s\tau)}{s^4} \frac{d\tau}{\tau^4}$$

$$+ \left[\frac{z^3}{6} \frac{K_0(sa/c_0)}{s} + \frac{azc_0}{2} \frac{K_1(sa/c_0)}{s^2} \right] H(z) \qquad (C.6)$$

which is further transformed to the TD with the aid of [25, eqs. (29.2.15) and (29.3.119)], that is

$$I^A(z,t) \simeq \frac{|z|^3}{12} \left\{ -\frac{11}{6} - \ln\left(\frac{c_0 t}{|z|}\right) + 3 \frac{c_0 t}{|z|} - \frac{3}{2}\left(\frac{c_0 t}{|z|}\right)^2 \right.$$

$$\left. + \frac{1}{3}\left(\frac{c_0 t}{|z|}\right)^3 \right\} H(t - |z|/c_0)$$

$$+ \frac{z}{6}\left(z^2 - \frac{3}{2}a^2\right)\cosh^{-1}\left(\frac{c_0 t}{a}\right) H(t - a/c_0)H(z)$$

$$+ \frac{1}{4}c_0 taz\left(\frac{c_0^2 t^2}{a^2} - 1\right)^{1/2} H(t - a/c_0)H(z) \qquad (C.7)$$

This expression will be used to derive the elements of the TD impedance matrix.

C.2   GENERIC INTEGRAL I^B

Referring again to Eqs. (C.1) and (C.2), the second generic integral has the form as given here

$$\hat{I}^B(z,s) = \frac{c_0^2}{2i\pi} \int_{p\in\mathbb{I}_0} \frac{\exp(spz)}{s^4 p^2} K_0[s\gamma_0(p)a]dp \qquad (C.8)$$

for $z \in \mathbb{R}$ and $\{s \in \mathbb{R}; s > 0\}$. Following the same lines of reasoning as in the preceding section, we first get

$$\hat{I}^B(z,s) = \frac{c_0^2|z|}{\pi} \int_{\tau=|z|/c_0}^{\infty} \frac{\exp(-s\tau)}{s^4} \mathrm{Im}\{K_0[-isa(\tau^2/z^2 - 1/c_0^2)^{1/2}]\}\frac{d\tau}{\tau^2}$$

$$+ \frac{c_0^2 z}{s^3} K_0(sa/c_0)H(z) \qquad (C.9)$$

For a relatively thin wire, the integral can be, again, approximated by

$$\hat{I}^B(z,s) \simeq \frac{c_0^2|z|}{2} \int_{\tau=|z|/c_0}^{\infty} \frac{\exp(-s\tau)}{s^4} \frac{d\tau}{\tau^2} + \frac{c_0^2 z}{s^3} K_0(sa/c_0)H(z) \qquad (C.10)$$

that can be readily transformed to the TD with the help of [25, eqs. (29.2.15) and (29.3.119)]. This way yields

$$I^B(z,t) \simeq \frac{|z|^3}{12} \left\{ \frac{1}{2} - 3\left(\frac{c_0 t}{|z|}\right)^2 \ln\left(\frac{c_0 t}{|z|}\right) - 3\frac{c_0 t}{|z|} + \frac{3}{2}\left(\frac{c_0 t}{|z|}\right)^2 \right.$$
$$\left. + \left(\frac{c_0 t}{|z|}\right)^3 \right\} H(t - |z|/c_0)$$
$$+ \frac{z}{2}\left(c_0^2 t^2 - \frac{1}{2}a^2\right) \cosh^{-1}\left(\frac{c_0 t}{a}\right) H(t - a/c_0)H(z)$$
$$- \frac{3}{4}c_0 t a z \left(\frac{c_0^2 t^2}{a^2} - 1\right)^{1/2} H(t - a/c_0)H(z) \qquad \text{(C.11)}$$

This expression will be used to derive the TD counterpart of Eq. (C.1).

C.3 TD IMPEDANCE MATRIX ELEMENTS

Collecting the results, the TD impedance matrix constituents can be found from

$$Z^{[S,n]}(t) = \frac{\zeta_0/2\pi}{c_0 \Delta t \Delta^2}[I(z_S - z_n + 2\Delta, t) - 4I(z_S - z_n + \Delta, t)$$
$$+ 6I(z_S - z_n, t) - 4I(z_S - z_n - \Delta, t) + I(z_S - z_n - 2\Delta, t)] \qquad \text{(C.12)}$$

in which

$$I(z,t) = I^B(z,t) - I^A(z,t)$$
$$\simeq \frac{|z|^3}{12} \left\{ \frac{7}{3} + \ln\left(\frac{c_0 t}{|z|}\right) - 3\left(\frac{c_0 t}{|z|}\right)^2 \ln\left(\frac{c_0 t}{|z|}\right) - 6\frac{c_0 t}{|z|} \right.$$
$$\left. + 3\left(\frac{c_0 t}{|z|}\right)^2 + \frac{2}{3}\left(\frac{c_0 t}{|z|}\right)^3 \right\} H(c_0 t - |z|) - c_0^2 t^2 z H(z)$$
$$+ \frac{z}{2}\left(c_0^2 t^2 - \frac{z^2}{3}\right) \cosh^{-1}\left(\frac{c_0 t}{a}\right) H(c_0 t - a)H(z) \qquad \text{(C.13)}$$

where we have neglected terms $O(a^2)$ as $a \downarrow 0$. Via Eqs. (C.12) and (C.13) we specified the elements of the TD impedance matrix introduced in section 2.3.

APPENDIX D

MUTUAL-IMPEDANCE MATRIX

The elements of the mutual-impedance matrix elements as appear in section 3.3 can be represented through (cf. Eq. (C.1))

$$
\hat{Z}_M^{[S,n]}(r, s) = \frac{\zeta_0/2\pi}{c_0 \Delta t \Delta^2} \left\{ \frac{c_0^2}{2i\pi} \int_{p=-i\infty}^{i\infty} \frac{\tilde{E}^{[S,n]}(p, s)}{s^4 p^2} K_0[s\gamma_0(p)r] dp \right.
$$

$$
\left. - \frac{1}{2i\pi} \int_{p=-i\infty}^{i\infty} \frac{\tilde{E}^{[S,n]}(p, s)}{s^4 p^4} K_0[s\gamma_0(p)r] dp \right\} \tag{D.1}
$$

for all $S = \{1, \ldots, N\}$ and $n = \{1, \ldots, N\}$, where $r > 0$ and $\zeta_0 = (\mu_0/\epsilon_0)^{1/2}$ is the free-space impedance and

$$
\tilde{E}^{[S,n]}(p, s) = [\exp(2sp\Delta) - 4\exp(sp\Delta) + 6 - 4\exp(-sp\Delta)
$$

$$
+ \exp(-2sp\Delta)] \exp[sp(z_S - z_n)] \tag{C.2 revisited}
$$

Following the approach used in appendix C, the integration path is first indented around the origin (see Figure C.1), and the mutual-impedance matrix elements are found upon evaluating generic integrals, handling of which is described in the ensuing sections.

D.1 GENERIC INTEGRAL J^A

Referring to Eqs. (D.1) and (C.2) given previously, the first generic integral has the following form

Time-Domain Electromagnetic Reciprocity in Antenna Modeling, First Edition. Martin Štumpf.
© 2020 by The Institute of Electrical and Electronics Engineers, Inc. Published 2020 by John Wiley & Sons, Inc.

$$\hat{J}^A(r, z, s) = \frac{1}{2i\pi} \int_{p \in \mathbb{I}_0} \frac{\exp(spz)}{s^4 p^4} K_0[s\gamma_0(p)r]dp \tag{D.2}$$

for $z \in \mathbb{R}$, $r > 0$ and $\{s \in \mathbb{R}; s > 0\}$, where the integration path is shown in Figure C.1. The modified Bessel function in Eq. (D.2) is next expressed via its integral representation [25, eq. (9.6.23)], which after changing the order of integration yields

$$\hat{J}^A(r, z, s) = \frac{1}{2i\pi} \int_{u=1}^{\infty} \frac{du}{(u^2 - 1)^{1/2}}$$
$$\int_{p \in \mathbb{I}_0} \exp\{-s[-pz + \gamma_0(p)ur]\} \frac{dp}{s^4 p^4} \tag{D.3}$$

In the complex p-plane, the integrand shows a fourfold pole singularity at the origin along with the branch cuts along $\{\text{Im}(p) = 0, 1/c_0 \le |\text{Re}(p)| < \infty\}$ due to $\gamma_0(p)$. Following the CdH methodology [8], the integration contour $\mathbb{I}_0$ is next replaced by the new CdH path that is defined via (cf. Eq. (C.5))

$$\tau = -pz + \gamma_0(p)ur \tag{D.4}$$

for $\{\tau \in \mathbb{R}; T(u) \le \tau < \infty\}$, where $T(u) = R(u)/c_0 = (z^2 + u^2 r^2)^{1/2}/c_0 > 0$, which is permissible thanks to Cauchy's theorem and Jordan's lemma. If $z > 0$, we must also add the contribution from the pole at $p = 0$. In this way, combining further the integrations in the upper and lower halves of the complex p-plane, we arrive at

$$\hat{J}^A(r, z, s) = \frac{1}{\pi} \int_{u=1}^{\infty} \frac{du}{(u^2 - 1)^{1/2}}$$
$$\int_{\tau=T(u)}^{\infty} \exp(-s\tau) \text{Re}\left\{\frac{\gamma_0[p(\tau)]}{s^4 p^4(\tau)}\right\} \frac{d\tau}{[\tau^2 - T^2(u)]^{1/2}}$$
$$+ \left[\frac{z^3}{6} \frac{K_0(sr/c_0)}{s} + \frac{rzc_0}{2} \frac{K_1(sr/c_0)}{s^2}\right] H(z) \tag{D.5}$$

where we take

$$p(\tau) = -\frac{z\tau}{c_0^2 T^2(u)} + \frac{iur}{c_0^2 T^2(u)}[\tau^2 - T^2(u)]^{1/2} \tag{D.6}$$

$$\gamma_0[p(\tau)] = \frac{ur\tau}{c_0^2 T^2(u)} + \frac{iz}{c_0^2 T^2(u)}[\tau^2 - T^2(u)]^{1/2} \tag{D.7}$$

Subsequently, interchanging the order of the integrations in Eq. (D.5), we get

$$\hat{J}^A(r, z, s) = \frac{1}{\pi} \frac{c_0}{s^4 r} \int_{\tau=T(1)}^{\infty} \exp(-s\tau) \mathrm{d}\tau$$

$$\int_{u=1}^{Q(\tau)} \mathrm{Re}\left\{ \frac{\gamma_0[p(\tau)]}{p^4(\tau)} \right\} \frac{\mathrm{d}u}{(u^2-1)^{1/2}[Q^2(\tau) - u^2]^{1/2}}$$

$$+ \left[\frac{z^3}{6} \frac{\mathrm{K}_0(sr/c_0)}{s} + \frac{rzc_0}{2} \frac{\mathrm{K}_1(sr/c_0)}{s^2} \right] \mathrm{H}(z) \qquad (\mathrm{D.8})$$

where we introduced $Q(\tau) = (c_0^2\tau^2 - z^2)^{1/2}/r > 0$. In the final step, the inner integral is handled using Eq. (A.9), and the result is transformed to the TD, which yields

$$J^A(r, z, t) = \frac{1}{6\pi r} \int_{v=R}^{c_0 t} (c_0 t - v)^3 \mathrm{d}v$$

$$\int_{\psi=0}^{\pi/2} \mathrm{Re}\left\{ \frac{c_0\overline{\gamma}_0[\overline{p}(v, \psi)]}{[c_0\overline{p}(v, \psi)]^4} \right\} \frac{\mathrm{d}\psi}{[\cos^2(\psi) + \overline{Q}^2(v)\sin^2(\psi)]^{1/2}}$$

$$+ z(z^2/6 - r^2/4) \cosh^{-1}(c_0 t/r) \mathrm{H}(t - r/c_0)\mathrm{H}(z)$$

$$+ (c_0 trz/4)(c_0^2 t^2/r^2 - 1)^{1/2}\mathrm{H}(t - r/c_0)\mathrm{H}(z) \qquad (\mathrm{D.9})$$

in which $\overline{Q}(v) = (v^2 - z^2)^{1/2}/r > 0$, $R = R(1) = (z^2 + r^2)^{1/2}$, and (cf. Eqs. (D.6) and (D.7))

$$c_0\overline{p}(v, \psi) = -\frac{zv}{R^2\cos^2(\psi) + v^2\sin^2(\psi)}$$

$$+ \frac{\mathrm{i}(v^2 - R^2)^{1/2}[r^2\cos^2(\psi) + (v^2 - z^2)\sin^2(\psi)]^{1/2}}{R^2\cos^2(\psi) + v^2\sin^2(\psi)} \cos(\psi) \quad (\mathrm{D.10})$$

$$c_0\overline{\gamma}_0[\overline{p}(v, \psi)] = \frac{[r^2\cos^2(\psi) + (v^2 - z^2)\sin^2(\psi)]^{1/2}v}{R^2\cos^2(\psi) + v^2\sin^2(\psi)}$$

$$+ \frac{\mathrm{i}(v^2 - R^2)^{1/2}z}{R^2\cos^2(\psi) + v^2\sin^2(\psi)} \cos(\psi) \qquad (\mathrm{D.11})$$

These relations will be used later in the expression for the TD mutual impedance.

D.2 GENERIC INTEGRAL J^B

The second generic integral, from which the mutual-impedance matrix can be constructed, reads

$$\hat{J}^B(r, z, s) = \frac{c_0^2}{2\mathrm{i}\pi} \int_{p\in\mathbb{I}_0} \frac{\exp(spz)}{s^4 p^2} \mathrm{K}_0[s\gamma_0(p)r]\mathrm{d}p \qquad (\mathrm{D.12})$$

for $z \in \mathbb{R}$, $r > 0$, and $\{s \in \mathbb{R}; s > 0\}$. Following the lines of reasoning closely described in the preceding section, we find the TD counterpart of the latter

$$J^B(r, z, t) = \frac{1}{6\pi r} \int_{v=R}^{c_0 t} (c_0 t - v)^3 dv$$
$$\int_{\psi=0}^{\pi/2} \text{Re}\left\{\frac{c_0 \overline{\gamma}_0[\overline{p}(v, \psi)]}{[c_0 \overline{p}(v, \psi)]^2}\right\} \frac{d\psi}{[\cos^2(\psi) + \overline{Q}^2(v)\sin^2(\psi)]^{1/2}}$$
$$+ z(c_0^2 t^2/2 + r^2/4)\cosh^{-1}(c_0 t/r)H(t - r/c_0)H(z)$$
$$- (3c_0 t r z/4)(c_0^2 t^2/r^2 - 1)^{1/2}H(t - r/c_0)H(z) \tag{D.13}$$

where $c_0\overline{p}(v, \psi)$ and $c_0\overline{\gamma}_0[\overline{p}(v, \psi)]$ are given in Eqs. (D.10) and (D.11).

D.3   TD MUTUAL-IMPEDANCE MATRIX ELEMENTS

Collecting the results, the TD mutual-impedance matrix constituents can be found from

$$Z_M^{[S,n]}(r, t) = \frac{\zeta_0/2\pi}{c_0 \Delta t \Delta^2}\left[J(r, z_S - z_n + 2\Delta, t) - 4J(r, z_S - z_n + \Delta, t)\right.$$
$$+ 6J(r, z_S - z_n, t) - 4J(r, z_S - z_n - \Delta, t)$$
$$\left.+ J(r, z_S - z_n - 2\Delta, t)\right] \tag{D.14}$$

in which

$$J(r, z, t) = J^B(r, z, t) - J^A(r, z, t)$$
$$= \frac{1}{6\pi} \int_{v=R}^{c_0 t} (c_0 t - v)^3 dv \int_{\psi=0}^{\pi/2} \text{Re}\left\{\frac{c_0 \overline{\gamma}_0[\overline{p}(v, \psi)]}{[c_0 \overline{p}(v, \psi)]^2}\right.$$
$$\left.- \frac{c_0 \overline{\gamma}_0[\overline{p}(v, \psi)]}{[c_0 \overline{p}(v, \psi)]^4}\right\} \frac{d\psi}{[r^2\cos^2(\psi) + (v^2 - z^2)\sin^2(\psi)]^{1/2}}$$
$$+ z(c_0^2 t^2/2 + r^2/2 - z^2/6)\cosh^{-1}(c_0 t/r)H(t - r/c_0)H(z)$$
$$- c_0 t r z(c_0^2 t^2/r^2 - 1)^{1/2}H(t - r/c_0)H(z) \tag{D.15}$$

where $c_0\overline{p}(v, \psi)$ and $c_0\overline{\gamma}_0[\overline{p}(v, \psi)]$ are given via Eqs. (D.10) and (D.11). In numerical implementations, we may further stretch the variable of the integration with respect to v using

$$v = R\cosh(u) \tag{D.16}$$

for $\{0 \le u \le \cosh^{-1}(c_0 t/R) < \infty\}$.

Finally note that the impedance matrix described in appendix C can be, in fact, viewed as a limiting case of the mutual-impedance matrix, that is

$$Z^{[S,n]}(t) = \lim_{r=a\downarrow 0} Z_M^{[S,n]}(r,t) \tag{D.17}$$

for which the integrations involved have been evaluated analytically in an approximate way. When calculating the pulsed EM mutual coupling between wire antennas, r has the meaning of the horizontal offset D between two interacting antennas (see Figure 3.1).

APPENDIX E

INTERNAL IMPEDANCE OF A SOLID WIRE

In appendix B, we have analyzed the EM radiation from a gap-excited, infinitely-long PEC wire antenna. If the radiating wire is not perfectly electrically conducting, then the jump boundary condition (B.4) has to account for the electric field inside the cylinder. Therefore, we write

$$\lim_{r\downarrow a} \hat{E}_z^s(r, z, s) - \lim_{r\uparrow a} \hat{E}_z^s(r, z, s) = -\hat{V}^T(s)\delta(z) \tag{E.1}$$

for all $z \in \mathbb{R}$. Consequently, referring to the slowness representation (A.1), we look for the solution in the following form

$$\hat{E}_z^s(r, z, s) = \frac{s\hat{V}^T(s)}{2\pi i} \int_{p=-i\infty}^{i\infty} \tilde{B}(p, s)I_0[s\hat{\gamma}_1(p, s)r]\exp(-spz)dp \tag{E.2}$$

for $\{0 < r < a\}$, where $\hat{\gamma}_1$ denotes the radial slowness parameter inside the cylinder. For a homogeneous and isotropic cylinder whose EM properties are described by electric permittivity ϵ and electric conductivity σ, the slowness parameter has the following form

$$\hat{\gamma}_1(p, s) = [\,\mu_0(\epsilon + \sigma/s) - p^2\,]^{1/2} \tag{E.3}$$

Exterior to the cylindrical antenna, the complex FD representation of the axial electric field decays exponentially with the increasing radial distance $r > 0$ from the antenna structure, that is

$$\hat{E}_z^s(r, z, s) = \frac{s\hat{V}^T(s)}{2\pi i} \int_{p=-i\infty}^{i\infty} \tilde{A}(p, s)K_0[s\gamma_0(p)r]\exp(-spz)dp \quad \text{(B.5 revisited)}$$

Time-Domain Electromagnetic Reciprocity in Antenna Modeling, First Edition. Martin Štumpf.
© 2020 by The Institute of Electrical and Electronics Engineers, Inc. Published 2020 by John Wiley & Sons, Inc.

for $r > a$. The unknown coefficients $\{\tilde{A}, \tilde{B}\}(p, s)$ are next found with the aid of Eq. (E.1) and the continuity-type boundary condition applying to the azimuthal component of the magnetic-field strength, that is

$$\lim_{r \downarrow a} \hat{H}_\phi^s(r, z, s) - \lim_{r \uparrow a} \hat{H}_\phi^s(r, z, s) = 0 \tag{E.4}$$

for all $z \in \mathbb{R}$. In this way, after straightforward algebra, we end up with

$$\tilde{B}(p, s) = \frac{\hat{\eta}_0/\hat{\eta}}{I_1(s\hat{\gamma}_1 a)} \left[\frac{\gamma_0}{\hat{\gamma}_1} \frac{K_0(s\gamma_0 a)}{K_1(s\gamma_0 a)} + \frac{\hat{\eta}_0}{\hat{\eta}} \frac{I_0(s\hat{\gamma}_1 a)}{I_1(s\hat{\gamma}_1 a)} \right]^{-1} \tag{E.5}$$

$$\tilde{A}(p, s) = \frac{-1}{K_0(s\gamma_0 a)} \frac{\gamma_0}{\hat{\gamma}_1} \frac{K_0(s\gamma_0 a)}{K_1(s\gamma_0 a)} \left[\frac{\gamma_0}{\hat{\gamma}_1} \frac{K_0(s\gamma_0 a)}{K_1(s\gamma_0 a)} + \frac{\hat{\eta}_0}{\hat{\eta}} \frac{I_0(s\hat{\gamma}_1 a)}{I_1(s\hat{\gamma}_1 a)} \right]^{-1} \tag{E.6}$$

where $\hat{\eta}_0 = s\epsilon_0$ and $\hat{\eta} = s\epsilon + \sigma$. It is noted that as $|\hat{\eta}| \to \infty$, we get $\tilde{B} = 0$ and $\tilde{A} = -1/K_0(s\gamma_0 a)$, which agrees with Eq. (B.6) applying to the PEC case. Employing Eqs. (E.5) and (E.6), we may further interrelate the transform-domain axial electric-field strength in the cylinder with the corresponding azimuthal magnetic-field strength, that is

$$\lim_{r \uparrow a} \tilde{E}_z(r, p, s) = \tilde{Z}(p, s) \, 2\pi a \lim_{r \uparrow a} \tilde{H}_\phi(r, p, s) \tag{E.7}$$

via the transform-domain internal-impedance of the solid wire

$$\tilde{Z}(p, s) = \frac{1}{\hat{\eta}} \frac{s\hat{\gamma}_1}{2\pi a} \frac{I_0(s\hat{\gamma}_1 a)}{I_1(s\hat{\gamma}_1 a)} \tag{E.8}$$

For a cylindrical antenna showing a high contrast with respect to its embedding, the field variations along its axis sufficiently far away from the excitation gap can be essentially neglected, and we write

$$\hat{Z}(s) = \tilde{Z}(0, s) = \frac{1}{2\pi a} \left(\frac{s\mu_0}{\sigma} \right)^{1/2} \frac{I_0[(s\mu_0\sigma)^{1/2}a]}{I_1[(s\mu_0\sigma)^{1/2}a]} \tag{E.9}$$

where we have assumed that the contribution of conductive electric currents in the wire dominates. Moreover, whenever $|(s\mu_0\sigma)^{1/2}a| \to \infty$, the large-argument approximation of (the ratio of) the modified Bessel functions leads to

$$\hat{Z}(s) \simeq \frac{1}{2\pi a} \left(\frac{s\mu_0}{\sigma} \right)^{1/2} \tag{E.10}$$

which is the "high-frequency approximation" previously incorporated in a finite-difference computational scheme [53]. This result is also used in chapter 4 to analyze EM scattering from a straight wire antenna including the ohmic loss.

APPENDIX F

VED-INDUCED EM COUPLING TO TRANSMISSION LINES – GENERIC INTEGRALS

This section is the addendum to chapter 13, where the VED–induced voltage responses on a transmission line are expressed analytically in closed form. The main mathematical tool in our analysis is the CdH method [8].

F.1 GENERIC INTEGRAL $\mathcal{I}$

The first generic representation used for calculating the VED-induced voltages on a transmission line can be written as

$$\hat{\mathcal{I}}(x, y, z, s) = \frac{c_0}{8\pi^2 \mathrm{i}^2} \int_{\kappa=-\mathrm{i}\infty}^{\mathrm{i}\infty} \exp(-s\kappa x) \frac{\kappa}{\kappa + c_0^{-1}} \mathrm{d}\kappa$$

$$\times \int_{\sigma=-\mathrm{i}\infty}^{\mathrm{i}\infty} \exp\{-s[\sigma y + \Gamma_0(\kappa, \sigma)z]\} \mathrm{d}\sigma \qquad \text{(F.1)}$$

for $x \in \mathbb{R}$, $y \in \mathbb{R}$, $\{z \in \mathbb{R}; z > 0\}$, and $\{s \in \mathbb{R}; s > 0\}$, where the wave-slowness parameter in the z-direction is given by

$$\Gamma_0(\kappa, \sigma) = [\Omega_0^2(\kappa) - \sigma^2]^{1/2} \qquad \text{(F.2)}$$

with $\Omega_0(\kappa) \triangleq (1/c_0^2 - \kappa^2)^{1/2}$ and $\mathrm{Re}(\Omega_0) \geq 0$. In the first step, the integrand in the integral with respect to σ is continued analytically into the complex σ-plane such that $\mathrm{Re}[\Gamma_0(\kappa, \sigma)] \geq 0$ for all $\sigma \in \mathbb{C}$, thereby introducing the branch cuts along $\{\mathrm{Im}(\sigma) = 0, \Omega_0(\kappa) < |\mathrm{Re}(\sigma)| < \infty\}$. Upon applying Jordan's lemma and

Time-Domain Electromagnetic Reciprocity in Antenna Modeling, First Edition. Martin Štumpf.
© 2020 by The Institute of Electrical and Electronics Engineers, Inc. Published 2020 by John Wiley & Sons, Inc.

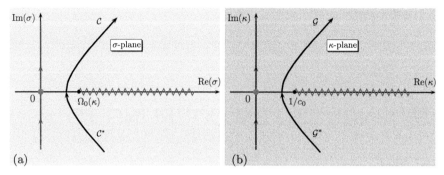

FIGURE F.1. (a) Complex σ-plane and (b) complex κ-plane and the new CdH paths for $y > 0$ and $x > 0$, respectively.

Cauchy's theorem, the original integration contour along $\text{Re}(\sigma) = 0$ is replaced with the CdH path, denoted by $\mathcal{C} \cup \mathcal{C}^*$, that follows from

$$\sigma y + \Gamma_0(\kappa, \sigma) z = ud\, \Omega_0(\kappa) \tag{F.3}$$

for $\{1 \leq u < \infty\}$ with $d \triangleq (y^2 + z^2)^{1/2} > 0$. Solving Eq. (F.3) for σ, we find

$$\mathcal{C} = \left\{ \sigma(u) = \left[yu/d + iz(u^2 - 1)^{1/2}/d \right] \Omega_0(\kappa) \right\} \tag{F.4}$$

for all $\{1 \leq u < \infty\}$ (see Figure F.1a). Along $\mathcal{C}$, we then have

$$\Gamma_0[\kappa, \sigma(u)] = \left[zu/d - iy(u^2 - 1)^{1/2}/d \right] \Omega_0(\kappa) \tag{F.5}$$

while the Jacobian of the mapping reads

$$\partial\sigma/\partial u = i\Gamma_0[\kappa, \sigma(u)]/(u^2 - 1)^{1/2} \tag{F.6}$$

for all $\{1 \leq u < \infty\}$. Upon combining the contributions from $\mathcal{C}$ and $\mathcal{C}^*$ and changing the order of the integrations, we end up with

$$\hat{\mathcal{I}}(x, y, z, s) = \frac{c_0}{4\pi^2 i} \frac{z}{d} \int_{u=1}^{\infty} \frac{u\, du}{(u^2 - 1)^{1/2}}$$
$$\times \int_{\kappa=-i\infty}^{i\infty} \exp\{-s[\kappa x + ud\, \Omega_0(\kappa)]\} \left(\frac{c_0^{-1} - \kappa}{c_0^{-1} + \kappa} \right)^{1/2} \kappa d\kappa \tag{F.7}$$

We shall next proceed in a similar way in the complex κ-plane. Hence, the integrand is first continued analytically away from the imaginary axis while keeping $\text{Re}[\Omega_0(\kappa)] \geq 0$ throughout the complex κ-plane, where we encounter branch points at $\kappa = \pm c_0^{-1}$ and the corresponding branch cuts along $\{\text{Im}(\kappa) = 0, c_0^{-1} <$

$|\mathrm{Re}(\kappa)| < \infty\}$. Subsequently, in virtue of Jordan's lemma and Cauchy's theorem, again, the original contour along $\mathrm{Re}(\kappa) = 0$ is deformed into the new CdH path, $\mathcal{G} \cup \mathcal{G}^*$, that is defined via

$$\kappa x + ud\, \Omega_0(\kappa) = \tau \tag{F.8}$$

for $\{\tau \in \mathbb{R}; \tau > 0\}$. Solving Eq. (F.8) for κ, we obtain

$$\mathcal{G} = \left\{ \kappa(\tau) = [x/R^2(u)]\tau + \mathrm{i}[ud/R^2(u)][\tau^2 - R^2(u)/c_0^2]^{1/2} \right\} \tag{F.9}$$

for $\tau \geq R(u)/c_0$, where we defined $R(u) \triangleq (x^2 + u^2 d^2)^{1/2} > 0$ (see Figure F.1b). Along $\mathcal{G}$, we then have

$$\Omega_0[\kappa(\tau)] = [ud/R^2(u)]\tau - \mathrm{i}[x/R^2(u)][\tau^2 - R^2(u)/c_0^2]^{1/2} \tag{F.10}$$

and

$$\partial\kappa/\partial\tau = \mathrm{i}\Omega_0[\kappa(\tau)]/[\tau^2 - R^2(u)/c_0^2]^{1/2} \tag{F.11}$$

for all $\tau \geq R(u)/c_0$. After combining the contributions of the integrations from $\mathcal{G} \cup \mathcal{G}^*$, we may rewrite Eq. (F.7) to the following form

$$\hat{\mathcal{I}}(x, y, z, s) = \frac{c_0}{2\pi^2} \frac{z}{d} \int_{u=1}^{\infty} \frac{u\, du}{(u^2 - 1)^{1/2}}$$

$$\times \int_{\tau=R(u)/c_0}^{\infty} \exp(-s\tau) \frac{\mathrm{Re}\left\{\kappa(\tau)[c_0^{-1} - \kappa(\tau)]\right\}}{[\tau^2 - R^2(u)/c_0^2]^{1/2}} d\tau \tag{F.12}$$

with the values of $\kappa(\tau)$ taken along $\mathcal{G}$ (see Eq. (F.9)). Expressing the integrand in Eq. (F.12) explicitly and interchanging the order of the integrations, we obtain (cf. Eq. (A.8))

$$\hat{\mathcal{I}}(x, y, z, s) = \frac{1}{2\pi^2} \frac{z}{d^2} \int_{\tau=R/c_0}^{\infty} \exp(-s\tau) d\tau$$

$$\times \int_{u=1}^{Q(\tau)} \left[\frac{xc_0\tau}{R^2(u)} - \frac{u^2 d^2}{R^2(u)} - \frac{x^2 c_0^2 \tau^2}{R^4(u)} + \frac{u^2 d^2 c_0^2 \tau^2}{R^4(u)} \right]$$

$$\times \frac{u\, du}{[Q^2(\tau) - u^2]^{1/2}(u^2 - 1)^{1/2}} \tag{F.13}$$

where $Q(\tau) = (c_0^2 \tau^2 - x^2)^{1/2}/d > 0$ and $R = R(x, y, z) = R(1) = (x^2 + y^2 + z^2)^{1/2} > 0$. The inner integrals with respect to u will be next evaluated analytically. With the help of the substitution defined by Eq. (A.9), the integrals can be rewritten as

$$
xc_0\tau \int_{\psi=0}^{\pi/2} \frac{\mathrm{d}\psi}{R^2\cos^2(\psi) + c_0^2\tau^2\sin^2(\psi)}
$$

$$
- \int_{\psi=0}^{\pi/2} \frac{d^2\cos^2(\psi) + (c_0^2\tau^2 - x^2)\sin^2(\psi)}{R^2\cos^2(\psi) + c_0^2\tau^2\sin^2(\psi)}\mathrm{d}\psi
$$

$$
- x^2 c_0^2\tau^2 \int_{\psi=0}^{\pi/2} \frac{\mathrm{d}\psi}{[R^2\cos^2(\psi) + c_0^2\tau^2\sin^2(\psi)]^2}
$$

$$
+ c_0^2\tau^2 \int_{\psi=0}^{\pi/2} \frac{d^2\cos^2(\psi) + (c_0^2\tau^2 - x^2)\sin^2(\psi)}{[R^2\cos^2(\psi) + c_0^2\tau^2\sin^2(\psi)]^2}\mathrm{d}\psi \tag{F.14}
$$

The latter integrals are next handled with the aid of standard integral formulas, namely

$$
\int_{\psi=0}^{\pi/2} \frac{D^2\cos^2(\psi) + C^2\sin^2(\psi)}{A^2\cos^2(\psi) + B^2\sin^2(\psi)}\mathrm{d}\psi = \frac{\pi}{2}\frac{AC^2 + BD^2}{AB(A + B)} \tag{F.15}
$$

$$
\int_{\psi=0}^{\pi/2} \frac{D^2\cos^2(\psi) + C^2\sin^2(\psi)}{[A^2\cos^2(\psi) + B^2\sin^2(\psi)]^2}\mathrm{d}\psi = \frac{\pi}{4}\frac{A^2C^2 + B^2D^2}{A^3B^3} \tag{F.16}
$$

Subsequently, applying Eqs. (F.15) and (F.16) to Eq. (F.14) and substituting the result in Eq. (F.13), we end up with

$$
\hat{\mathcal{I}}(x, y, z, s) = (z/4\pi d^2) \int_{\tau=R/c_0}^{\infty} \exp(-s\tau)\mathcal{P}(x, y, z, \tau)\mathrm{d}\tau \tag{F.17}
$$

in which

$$
\mathcal{P}(x, y, z, \tau) = (1/Rc_0\tau)
$$

$$
\left[xc_0\tau - x^2 - \frac{R(c_0^2\tau^2 - x^2) + c_0\tau(y^2 + z^2)}{R + c_0\tau} + \frac{c_0^2\tau^2(y^2 + z^2)}{R^2}\right] \tag{F.18}
$$

and recall that $R = R(x, y, z)$. The TD counterpart of Eq. (F.17) follows upon employing Lerch's uniqueness theorem of the one-sided Laplace transformation [4, appendix]. In this way, we finally obtain

$$
\mathcal{I}(x, y, z, t) = [z/4\pi(y^2 + z^2)]
$$

$$
\times \mathcal{P}(x, y, z, t)\mathrm{H}[t - R(x, y, z)/c_0] \tag{F.19}
$$

where $\mathrm{H}(t)$ denotes the Heaviside unit-step function. This result is used in section 13.1.

F.2 GENERIC INTEGRAL $\mathcal{J}$

The second generic integral to be handled via the CdH method has the following form

$$\hat{\mathcal{J}}(x,y,z,s) = \frac{c_0}{8\pi^2 i^2} \int_{\kappa=-i\infty}^{i\infty} d\kappa$$

$$\times \int_{\sigma=-i\infty}^{i\infty} \exp\{-s[\kappa x + \sigma y + \Gamma_0(\kappa,\sigma)z]\} \frac{\kappa^2 + \sigma^2}{\Gamma_0^2(\kappa,\sigma)} d\sigma \quad \text{(F.20)}$$

for $x \in \mathbb{R}$, $y \in \mathbb{R}$, $\{z \in \mathbb{R}; z > 0\}$, and $\{s \in \mathbb{R}; s > 0\}$. To find the TD counterpart of $\hat{\mathcal{J}}(x,y,z,s)$, we first replace the variables of integration in Eq. (F.20) by $\{v,q\}$ via the following transformation

$$\kappa = v\cos(\phi) - iq\sin(\phi) \tag{F.21}$$

$$\sigma = v\sin(\phi) + iq\cos(\phi) \tag{F.22}$$

with $x = r\cos(\phi)$ and $y = r\sin(\phi)$ for $r \geq 0$ and $\{0 \leq \phi < 2\pi\}$. Under the transformation, $\kappa x + \sigma y = vr$, $\kappa^2 + \sigma^2 = v^2 - q^2$, and $d\kappa d\sigma = idvdq$. Consequently, the generic integral transforms to

$$\hat{\mathcal{J}}(x,y,z,s) = \frac{c_0}{8\pi^2 i} \int_{q=-\infty}^{\infty} dq$$

$$\times \int_{v=-i\infty}^{i\infty} \exp\{-s[vr + \overline{\Gamma}_0(v,q)z]\} \frac{v^2 - q^2}{\overline{\Gamma}_0^2(v,q)} dv \quad \text{(F.23)}$$

where (cf. Eq. (F.2))

$$\overline{\Gamma}_0(v,q) = [\overline{\Omega}_0^2(q) - v^2]^{1/2} \tag{F.24}$$

with $\overline{\Omega}_0(q) \triangleq (1/c_0^2 + q^2)^{1/2} > 0$. Next, the integrand in the integral with respect to v is analytically continued into the complex v-plane, away from the imaginary v-axis. In this process, we keep $\text{Re}[\overline{\Gamma}_0(v,q)] \geq 0$, thus introducing the branch cuts along $\{\text{Im}(v) = 0, \overline{\Omega}_0(q) < |\text{Re}(v)| < \infty\}$. Subsequently, with the aid of Jordan's lemma and Cauchy's theorem, the original integration contour is deformed into the CdH path, $\mathcal{E} \cup \mathcal{E}^*$, that is defined via

$$vr + \overline{\Gamma}_0(v,q)z = \tau \tag{F.25}$$

for $\{\tau \in \mathbb{R}; \tau > 0\}$. Solving then Eq. (F.25) for v, we get

$$\mathcal{E} = \left\{ v(\tau) = (r/R^2)\tau + i(z/R^2)[\tau^2 - R^2\,\overline{\Omega}_0^2(q)]^{1/2} \right\} \tag{F.26}$$

for $\tau \geq R\,\overline{\Omega}_0(q)$, where $R = (r^2 + z^2)^{1/2} = (x^2 + y^2 + z^2)^{1/2} > 0$. Along $\mathcal{E}$ in $\mathrm{Im}(v) > 0$, we have

$$\overline{\Gamma}_0[v(\tau), q] = (z/R^2)\tau - \mathrm{i}(r/R^2)[\tau^2 - R^2\,\overline{\Omega}_0^2(q)]^{1/2} \tag{F.27}$$

and

$$\partial v/\partial \tau = \mathrm{i}\overline{\Gamma}_0[v(\tau), q]/[\tau^2 - R^2\,\overline{\Omega}_0^2(q)]^{1/2} \tag{F.28}$$

for all $\tau \geq R\,\overline{\Omega}_0(q)$. Upon combining the integrations along $\mathcal{E}$ and $\mathcal{E}^*$, we then find

$$\hat{\mathcal{J}}(x, y, z, s) = \frac{c_0}{4\pi^2} \int_{q=-\infty}^{\infty} \mathrm{d}q$$
$$\times \int_{\tau=R\,\overline{\Omega}_0(q)}^{\infty} \exp(-s\tau)\mathrm{Re}\left\{\frac{v^2(\tau) - q^2}{\overline{\Gamma}_0[v(\tau), q]}\right\} \frac{\mathrm{d}\tau}{[\tau^2 - R^2\,\overline{\Omega}_0^2(q)]^{1/2}} \tag{F.29}$$

The change of the order of the integrations in Eq. (F.29) leads to

$$\hat{\mathcal{J}}(x, y, z, s) = \frac{c_0}{4\pi^2 R} \int_{\tau=R/c_0}^{\infty} \exp(-s\tau)\mathrm{d}\tau$$
$$\times \int_{q=-Q(\tau)}^{Q(\tau)} \mathrm{Re}\left\{\frac{v^2(\tau) - q^2}{\overline{\Gamma}_0[v(\tau), q]}\right\} \frac{\mathrm{d}q}{[Q^2(\tau) - q^2]^{1/2}} \tag{F.30}$$

where we used $Q(\tau) = (\tau^2/R^2 - 1/c_0^2)^{1/2}$. Rewriting the integrand with respect to q to its explicit form, we get

$$\hat{\mathcal{J}}(x, y, z, s) = \frac{c_0}{4\pi^2} \frac{z}{R} \int_{\tau=R/c_0}^{\infty} \exp(-s\tau)\tau\mathrm{d}\tau$$
$$\times \int_{q=-Q(\tau)}^{Q(\tau)} \frac{1}{c_0^2\tau^2 - r^2 - r^2 c_0^2 q^2} \frac{\mathrm{d}q}{[Q^2(\tau) - q^2]^{1/2}}$$
$$- \frac{c_0}{4\pi^2} \frac{z}{R^3} \int_{\tau=R/c_0}^{\infty} \exp(-s\tau)\tau\mathrm{d}\tau \int_{q=-Q(\tau)}^{Q(\tau)} \frac{\mathrm{d}q}{[Q^2(\tau) - q^2]^{1/2}} \tag{F.31}$$

The inner integrals with respect to q can be carried out analytically. Hence, we substitute

$$q = Q(\tau)\sin(u) \tag{F.32}$$

for $\{-\pi/2 \leq u \leq \pi/2\}$ and use a standard integral for the first integration, namely

$$\int_{u=-\pi/2}^{\pi/2} \frac{\mathrm{d}u}{1 - A^2\sin^2(u) - B^2\cos^2(u)} = \frac{\pi}{(1 - A^2)^{1/2}(1 - B^2)^{1/2}} \tag{F.33}$$

for $A^2 < 1$ and $B^2 < 1$. This way yields

$$\hat{\mathcal{J}}(x, y, z, s) = (1/4\pi) \int_{\tau = R/c_0}^{\infty} \exp(-s\tau)(c_0^2\tau^2 - r^2)^{-1/2} d\tau$$

$$- (c_0 z/4\pi R^3) \int_{\tau = R/c_0}^{\infty} \exp(-s\tau)\tau d\tau \qquad (\text{F.34})$$

where the integrals have the form of the one-sided Laplace transformation. Therefore, their TD counterparts follow upon inspection

$$\mathcal{J}(x, y, z, t) = (1/4\pi) \left[(c_0^2 t^2 - x^2 - y^2)^{-1/2} \right.$$

$$\left. - z\, c_0 t/R^3(x, y, z) \right] \mathrm{H}[t - R(x, y, z)/c_0] \qquad (\text{F.35})$$

relying, again, on Lerch's uniqueness theorem applying to the real-valued and positive transform parameter s. This result is also used in section 13.1.

F.3   GENERIC INTEGRAL $\mathcal{K}$

To evaluate the impact of a finite ground conductivity on the VED-induced Thévenin voltages on a transmission line via the Cooray-Rubinstein formula, we need to transform the following generic integral to the TD, that is

$$\hat{\mathcal{K}}(x, y, z, s) = \frac{1}{4\pi^2\mathrm{i}^2} \int_{\kappa = -\mathrm{i}\infty}^{\mathrm{i}\infty} \exp(-s\kappa x) \frac{\kappa}{\kappa + c_0^{-1}} d\kappa$$

$$\times \int_{\sigma = -\mathrm{i}\infty}^{\mathrm{i}\infty} \exp\{-s[\sigma y + \Gamma_0(\kappa, \sigma)z]\}\Gamma_0^{-1}(\kappa, \sigma) d\sigma \qquad (\text{F.36})$$

for $x \in \mathbb{R}$, $y \in \mathbb{R}$, $\{z \in \mathbb{R}; z > 0\}$, and $\{s \in \mathbb{R}; s > 0\}$, where $\Gamma_0(\kappa, \sigma)$ is defined by Eq. (F.2). Following the procedure described in section F.1, we deform the integration path in the complex σ-plane to the CdH path, $\mathcal{C} \cup \mathcal{C}^*$, as defined by Eq. (F.3) (see Figure F.1a). Upon combining the contributions from the integrations along $\mathcal{C}$ and $\mathcal{C}^*$ and changing the order of the integrations, we find

$$\hat{\mathcal{K}}(x, y, z, s) = \frac{1}{2\pi^2\mathrm{i}} \int_{u=1}^{\infty} \frac{du}{(u^2 - 1)^{1/2}}$$

$$\times \int_{\kappa = -\mathrm{i}\infty}^{\mathrm{i}\infty} \exp\{-s[\kappa x + ud\,\Omega_0(\kappa)]\} \frac{\kappa d\kappa}{\kappa + c_0^{-1}} \qquad (\text{F.37})$$

where we used Eq. (F.6). Since $d = (y^2 + z^2)^{1/2} > 0$ and $\mathrm{Re}[\Omega_0(\kappa)] \geq 0$ in the entire complex κ-plane, we may deform the integration contour in the complex

κ-plane to the CdH path $\mathcal{G} \cup \mathcal{G}^*$ that is defined by Eq. (F.8) (see Figure F.1b). In this way, we obtain

$$
\hat{\mathcal{K}}(x, y, z, s) = \frac{1}{\pi^2} \int_{u=1}^{\infty} \frac{du}{(u^2 - 1)^{1/2}}
$$
$$
\times \int_{\tau=R(u)/c_0}^{\infty} \exp(-s\tau) \mathrm{Re} \left\{ \frac{\kappa(\tau)\Omega_0[\kappa(\tau)]}{c_0^{-1} + \kappa(\tau)} \right\} \frac{d\tau}{[\tau^2 - R^2(u)/c_0^2]^{1/2}}
$$

$$(F.38)$$

In the ensuing step, we use Eqs. (F.9) and (F.10) to express the integrand with respect to τ explicitly. Moreover, we change the order of the integrations and get

$$
\hat{\mathcal{K}}(x, y, z, s) = \frac{1}{\pi^2 d} \int_{\tau=R/c_0}^{\infty} \exp(-s\tau) d\tau
$$
$$
\times \int_{u=1}^{Q(\tau)} \left[\frac{1}{R^2(u)} \frac{d(c_0^3\tau^3 + 2xc_0^2\tau^2)}{(c_0\tau + x)^2} - \frac{xd}{(c_0\tau + x)^2} - \frac{u^2}{R^2(u)} \frac{d^3 c_0\tau}{(c_0\tau + x)^2} \right]
$$
$$
\times \frac{u \, du}{[Q^2(\tau) - u^2]^{1/2}(u^2 - 1)^{1/2}}
$$

$$(F.39)$$

and recall that $Q(\tau) = (c_0^2\tau^2 - x^2)^{1/2}/d > 0$. The inner integrals with respect to u will be next rewritten via the substitution (A.9), which leads to

$$
\frac{d(c_0^3\tau^3 + 2xc_0^2\tau^2)}{(c_0\tau + x)^2} \int_{\psi=0}^{\pi/2} \frac{d\psi}{R^2\cos^2(\psi) + c_0^2\tau^2\sin^2(\psi)}
$$
$$
- \frac{xd}{(c_0\tau + x)^2} \int_{\psi=0}^{\pi/2} d\psi
$$
$$
- \frac{dc_0\tau}{(c_0\tau + x)^2} \int_{\psi=0}^{\pi/2} \frac{d^2\cos^2(\psi) + (c_0^2\tau^2 - x^2)\sin^2(\psi)}{R^2\cos^2(\psi) + c_0^2\tau^2\sin^2(\psi)} d\psi
$$

$$(F.40)$$

The latter integrals can be carried out with the aid of formulas (F.15) and (F.16), and we end up with

$$
\hat{\mathcal{K}}(x, y, z, s) = (1/2\pi R) \int_{\tau=R/c_0}^{\infty} \exp(-s\tau) \mathcal{H}(x, y, z, \tau) d\tau
$$

$$(F.41)$$

in which

$$
\mathcal{H}(x, y, z, \tau) = 1/(c_0\tau + x)^2
$$
$$
\times \left[c_0\tau(c_0\tau + 2x) - xR - \frac{R(c_0^2\tau^2 - x^2) + c_0\tau(y^2 + z^2)}{R + c_0\tau} \right]
$$

$$(F.42)$$

Upon noting that Eq. (F.41) has the form of the one-sided Laplace transformation, we employ Lerch's uniqueness theorem [4, appendix] and get

$$\mathcal{K}(x, y, z, t) = [1/2\pi R(x, y, z)]\mathcal{H}(x, y, z, t)\mathrm{H}[t - R(x, y, z)/c_0] \qquad \text{(F.43)}$$

These results are used in section 13.2 to evaluate the impact of the finite ground conductivity and permittivity on the VED-induced voltage response of a transmission line.

APPENDIX G

IMPEDITIVITY MATRIX

The TD impeditivity matrix elements are found from (cf. Eq. (C.12))

$$
\begin{aligned}
\mathcal{Z}^{[S,n]}(t) = [\zeta_0/c_0 \Delta t \Delta] \, [& R(x_S - x_n + 3\Delta/2, w/2, t) \\
& - 3R(x_S - x_n + \Delta/2, w/2, t) + 3R(x_S - x_n - \Delta/2, w/2, t) \\
& - R(x_S - x_n - 3\Delta/2, w/2, t)]
\end{aligned}
\tag{G.1}
$$

for all $S = \{1, \ldots, N\}$ and $n = \{1, \ldots, N\}$, where

$$
R(x, y, t) = J(x, y, t) - J(x, -y, t)
\tag{G.2}
$$

and $J(x, y, t)$ is obtained via the CdH method [8] applied to the generic integral representation as described in the next section.

G.1 GENERIC INTEGRAL J

The constituents of the complex FD impeditivity array can be composed of $\hat{J}(x, y, s)$ that can be expressed through the wave slowness representation (13.10) in the following form

$$
\begin{aligned}
\hat{J}(x, y, s) = \frac{c_0^2}{8\pi^2} \int_{\kappa \in \mathbb{K}_0} & \frac{\exp(s\kappa x)}{s^3 \kappa^3} \Omega_0^2(\kappa) \mathrm{d}\kappa \\
& \times \int_{\sigma \in \mathbb{S}_0} \frac{\exp(s\sigma y)}{s\sigma} \frac{\mathrm{d}\sigma}{\Gamma_0(\kappa, \sigma)}
\end{aligned}
\tag{G.3}
$$

for $\{x \in \mathbb{R}; x \neq 0\}$, $\{y \in \mathbb{R}; y \neq 0\}$, and $\{s \in \mathbb{R}; s > 0\}$, where $\mathbb{K}_0$ and $\mathbb{S}_0$ are the integrations contours running along the imaginary axes in the complex κ- and σ-planes, respectively, with the indentation around the origin as depicted in

Time-Domain Electromagnetic Reciprocity in Antenna Modeling, First Edition. Martin Štumpf.
© 2020 by The Institute of Electrical and Electronics Engineers, Inc. Published 2020 by John Wiley & Sons, Inc.

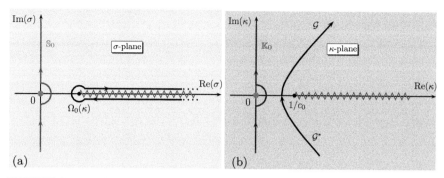

FIGURE G.1. (a) Complex σ-plane and (b) complex κ-plane with the original integration contours $\mathbb{S}_0$ and $\mathbb{K}_0$ and the new CdH paths for $y < 0$ and $x < 0$, respectively.

Figure G.1. Following the same line of reasoning as for the indentation of $\mathbb{I}_0$ in appendix C, such deformations do not change the result of the integrations in total, that is, in the representation for $\hat{Z}^{[S,n]}(s)$.

At first, the integrand in the integral with respect to σ is continued analytically into the complex σ-plane, while keeping $\mathrm{Re}[\Gamma_0(\kappa, \sigma)] \geq 0$. This implies the branch cuts along $\{\mathrm{Im}(\sigma) = 0, \Omega_0(\kappa) < |\mathrm{Re}(\sigma)| < \infty\}$. Subsequently, in virtue of Jordan's lemma and Cauchy's theorem, the integration path $\mathbb{S}_0$ is replaced with the loop encircling the branch cut. The loop is parametrized by

$$\sigma(u) = -u\Omega_0(\kappa)\mathrm{sgn}(y) \pm \mathrm{i}0 \qquad (\mathrm{G.4})$$

for $\{1 \leq u < \infty\}$, where $\mathrm{sgn}(y) = |y|/y$. Combining the integrations just above and just below the branch cut and including the simple pole contribution for $y > 0$, we find

$$\frac{1}{2\mathrm{i}\pi} \int_{\sigma \in \mathbb{S}_0} \frac{\exp(s\sigma y)}{s\sigma} \frac{\mathrm{d}\sigma}{\Gamma_0(\kappa, \sigma)} = \frac{\mathrm{H}(y)}{s\Omega_0(\kappa)}$$

$$- \frac{\mathrm{sgn}(y)}{s\Omega_0(\kappa)} \frac{1}{\pi} \int_{u=1}^{\infty} \exp[-su|y|\Omega_0(\kappa)] \frac{\mathrm{d}u}{u(u^2 - 1)^{1/2}} \qquad (\mathrm{G.5})$$

which is substituted back in Eq. (G.3). This way yields the sum of two integrals, say $\hat{J} = \hat{J}^A + \hat{J}^B$, where

$$\hat{J}^A(x, y, s) = \frac{\mathrm{sgn}(y)}{2\pi s} \int_{u=1}^{\infty} \frac{\mathrm{d}u}{u(u^2 - 1)^{1/2}}$$

$$\times \frac{c_0^2}{2\mathrm{i}\pi} \int_{\kappa \in \mathbb{K}_0} \exp\{-s[-\kappa x + u|y|\Omega_0(\kappa)]\} \frac{\Omega_0(\kappa)}{s^3 \kappa^3} \mathrm{d}\kappa \qquad (\mathrm{G.6})$$

after changing the order of integration and

$$\hat{j}^B(x, y, s) = -\frac{c_0^2}{2s} \frac{\mathrm{H}(y)}{2\mathrm{i}\pi} \int_{\kappa \in \mathbb{K}_0} \frac{\exp(s\kappa x)}{s^3 \kappa^3} \Omega_0(\kappa) \mathrm{d}\kappa \qquad (G.7)$$

The latter expressions will be next transformed to the TD.

G.1.1 Generic Integral J^A

We shall start with $\hat{J}^A(x, y, s)$ as given by Eq. (G.6). Here, the integrand with respect to κ is first continued analytically away from the imaginary axis, while keeping $\mathrm{Re}[\Omega_0(\kappa)] \geq 0$. This implies the branch cuts along the real κ-axis starting at the branch points $\pm 1/c_0$, that is, along $\{\mathrm{Im}(\kappa) = 0, c_0^{-1} < |\mathrm{Re}(\kappa)| < \infty\}$. Next, in virtue of Jordan's lemma and Cauchy's theorem, the original contour $\mathbb{K}_0$ is deformed into the CdH path, denoted by $\mathcal{G} \cup \mathcal{G}^*$, that is defined via (cf. Eq. (F.8))

$$-\kappa x + u|y| \, \Omega_0(\kappa) = \tau \qquad (G.8)$$

for $\{\tau \in \mathbb{R}; \tau > 0\}$. Solving the latter for κ, we obtain

$$\mathcal{G} = \left\{ \kappa(\tau) = [-x/r^2(u)]\tau + \mathrm{i}[u|y|/r^2(u)][\tau^2 - r^2(u)/c_0^2]^{1/2} \right\} \qquad (G.9)$$

for all $\tau \geq r(u)/c_0$, where we defined $r(u) \triangleq (x^2 + u^2 y^2)^{1/2} > 0$ (see Figure G.1b). Along $\mathcal{G}$, we then have

$$\Omega_0[\kappa(\tau)] = [u|y|/r^2(u)]\tau + \mathrm{i}[x/r^2(u)][\tau^2 - r^2(u)/c_0^2]^{1/2} \qquad (G.10)$$

while the Jacobian of the mapping reads here

$$\partial \kappa / \partial \tau = \mathrm{i}\Omega_0[\kappa(\tau)]/[\tau^2 - r^2(u)/c_0^2]^{1/2} \qquad (G.11)$$

for all $\tau \geq r(u)/c_0$. Combination of the contributions from $\mathcal{G}$ and $\mathcal{G}^*$ allows to rewrite Eq. (G.6) to the following form

$$\hat{j}^A(x, y, s) = \frac{c_0^3}{2\pi^2} \frac{\mathrm{sgn}(y)}{s^4} \int_{u=1}^{\infty} \frac{\mathrm{d}u}{u(u^2 - 1)^{1/2}} \int_{\tau=r(u)/c_0}^{\infty} \exp(-s\tau)$$

$$\times \mathrm{Re} \left\{ \frac{c_0^2 \Omega_0^2[\kappa(\tau)]}{c_0^3 \kappa^3(\tau)} \right\} \frac{\mathrm{d}\tau}{[\tau^2 - r^2(u)/c_0^2]^{1/2}} + \hat{j}^{AB} \qquad (G.12)$$

where we take the values along $\mathcal{G}$ according to Eqs. (G.9) and (G.10), and $\hat{J}^{AB}$ is the contribution originating from the triple pole singularity at $\kappa = 0$ for $x > 0$. Next, the integrand in the integral with respect to τ is expressed explicitly, and the order of the integrations is interchanged, which yields

$$
\hat{J}^A(x,y,s) = \frac{1}{2\pi^2} \frac{c_0^4}{s^4} \frac{1}{y} \int_{\tau=r/c_0}^{\infty} \exp(-s\tau)\mathrm{d}\tau \int_{u=1}^{Q(\tau)} \frac{1}{u^2}
$$

$$
\times \left[\frac{xc_0\tau}{c_0^2\tau^2 - u^2y^2} - \frac{x^3 c_0^3 \tau^3}{(c_0^2\tau^2 - u^2y^2)^3} - \frac{3x^3 c_0 \tau u^2 y^2}{(c_0^2\tau^2 - u^2y^2)^3} \right.
$$

$$
\left. + \frac{3xc_0\tau u^2 y^2}{(c_0^2\tau^2 - u^2y^2)^2} \right] \frac{u\,\mathrm{d}u}{[Q^2(\tau) - u^2]^{1/2}(u^2 - 1)^{1/2}} + \hat{J}^{AB} \quad \text{(G.13)}
$$

where we used $Q(\tau) = (c_0^2\tau^2 - x^2)^{1/2}/|y| > 0$ and $r = r(x,y) = r(1) = (x^2 + y^2)^{1/2} > 0$. The inner integrals with respect to u will be carried out analytically via the substitution (A.9). The latter yields the following form of the inner integrals

$$
\frac{xc_0\tau}{y^2} \int_{\psi=0}^{\pi/2} \frac{1}{\cos^2(\psi) + (c_0^2\tau^2/y^2 - x^2/y^2)\sin^2(\psi)}
$$

$$
\times \frac{\mathrm{d}\psi}{(c_0^2\tau^2/y^2 - 1)\cos^2(\psi) + (x^2/y^2)\sin^2(\psi)}
$$

$$
- \frac{x^3 c_0^3 \tau^3}{y^6} \int_{\psi=0}^{\pi/2} \frac{1}{\cos^2(\psi) + (c_0^2\tau^2/y^2 - x^2/y^2)\sin^2(\psi)}
$$

$$
\times \frac{\mathrm{d}\psi}{[(c_0^2\tau^2/y^2 - 1)\cos^2(\psi) + (x^2/y^2)\sin^2(\psi)]^3}
$$

$$
- \frac{3x^3 c_0 \tau}{y^4} \int_{\psi=0}^{\pi/2} \frac{\mathrm{d}\psi}{[(c_0^2\tau^2/y^2 - 1)\cos^2(\psi) + (x^2/y^2)\sin^2(\psi)]^3}
$$

$$
+ \frac{3xc_0\tau}{y^2} \int_{\psi=0}^{\pi/2} \frac{\mathrm{d}\psi}{[(c_0^2\tau^2/y^2 - 1)\cos^2(\psi) + (x^2/y^2)\sin^2(\psi)]^2} \quad \text{(G.14)}
$$

which will be evaluated with the aid of the following integral formulas

$$
\int_{\psi=0}^{\pi/2} \frac{1}{\cos^2(\psi) + (A^2 - B^2)\sin^2(\psi)} \frac{\mathrm{d}\psi}{(A^2 - 1)\cos^2(\psi) + B^2\sin^2(\psi)}
$$

$$
= \frac{\pi}{2A^2} \left[\frac{1}{(A^2 - B^2)^{1/2}} + \frac{1}{|B|(A^2 - 1)^{1/2}} \right] \quad \text{(G.15)}
$$

with $A^2 = c_0^2\tau^2/y^2$ and $B^2 = x^2/y^2$, and similarly

$$\int_{\psi=0}^{\pi/2} \frac{1}{\cos^2(\psi) + (A^2 - B^2)\sin^2(\psi)} \frac{d\psi}{[(A^2 - 1)\cos^2(\psi) + B^2\sin^2(\psi)]^3}$$

$$= \frac{\pi}{16} \frac{1}{A^6(A^2 - 1)^{5/2}|B|^5} [3A^8 + 6A^6(B^2 - 1) + A^4(15B^4 - 10B^2 + 3)$$

$$+ 4A^2B^2(1 - 5B^2) + 8B^4] + \frac{\pi}{2} \frac{1}{A^6(A^2 - B^2)^{1/2}} \tag{G.16}$$

$$\int_{\psi=0}^{\pi/2} \frac{d\psi}{[(A^2 - 1)\cos^2(\psi) + B^2\sin^2(\psi)]^3}$$

$$= \frac{\pi}{16} \frac{1}{(A^2 - 1)^{5/2}|B|^5} [3A^4 + 2A^2(B^2 - 3) + B^2(3B^2 - 2) + 3] \tag{G.17}$$

$$\int_{\psi=0}^{\pi/2} \frac{d\psi}{[(A^2 - 1)\cos^2(\psi) + B^2\sin^2(\psi)]^2}$$

$$= \frac{\pi}{4} \frac{1}{(A^2 - 1)^{3/2}|B|^3} [A^2 + B^2 - 1] \tag{G.18}$$

Hence, making use of Eqs. (G.15)–(G.18) in Eq. (G.13), we find

$$\hat{J}^A(x, y, s) = \frac{1}{2\pi} \frac{c_0^4}{s^4} \int_{\tau=r/c_0}^{\infty} \exp(-s\tau) F(x, y, \tau) d\tau + \hat{J}^{AB} \tag{G.19}$$

in which

$$F(x, y, \tau) = \frac{1}{2} \frac{\text{sgn}(x)\text{sgn}(y)}{c_0\tau} \left[\frac{1}{(c_0^2\tau^2/x^2 - 1)^{1/2}} + \frac{1}{(c_0^2\tau^2/y^2 - 1)^{1/2}} \right]$$

$$- \frac{1}{16} \frac{y^4}{x^2 c_0^3\tau^3} \frac{\text{sgn}(x)\text{sgn}(y)}{(c_0^2\tau^2/y^2 - 1)^{5/2}} \left[3\frac{c_0^8\tau^8}{y^8} + 6\frac{c_0^6\tau^6}{y^6} \left(\frac{x^2}{y^2} - 1 \right) \right.$$

$$\left. + \frac{c_0^4\tau^4}{y^4} \left(15\frac{x^4}{y^4} - 10\frac{x^2}{y^2} + 3 \right) + 4\frac{c_0^2\tau^2}{y^2} \frac{x^2}{y^2} \left(1 - 5\frac{x^2}{y^2} \right) + 8\frac{x^4}{y^4} \right]$$

$$- \frac{1}{2} \frac{x^2}{c_0^3\tau^3} \frac{\text{sgn}(x)\text{sgn}(y)}{(c_0^2\tau^2/x^2 - 1)^{1/2}} - \frac{3}{16} \frac{c_0\tau}{x^2} \frac{\text{sgn}(x)\text{sgn}(y)}{(c_0^2\tau^2/y^2 - 1)^{5/2}} \left[3\frac{c_0^4\tau^4}{y^4} \right.$$

$$\left. + 2\frac{c_0^2\tau^2}{y^2} \left(\frac{x^2}{y^2} - 3 \right) + \frac{x^2}{y^2} \left(3\frac{x^2}{y^2} - 2 \right) + 3 \right]$$

$$+ \frac{3}{4} \frac{c_0\tau}{x^2} \frac{\text{sgn}(x)\text{sgn}(y)}{(c_0^2\tau^2/y^2 - 1)^{3/2}} \left(\frac{c_0^2\tau^2}{y^2} + \frac{x^2}{y^2} - 1 \right) \tag{G.20}$$

It remains to specify $\hat{J}^{AB} = \hat{J}^{AB}(x, y, s)$. Accordingly, we start over from Eq. (G.6) where we evaluate the contribution from the triple pole singularity at the origin of κ-plane. Upon employing the formula of Cauchy [30, section 2.41], we find

$$
\hat{J}^{AB}(x, y, s) = \frac{c_0}{4\pi} \frac{H(x)\mathrm{sgn}(y)}{s^4} \left(s^2 x^2 - c_0^2\right) \int_{u=1}^{\infty} \frac{\exp(-su|y|/c_0)\mathrm{d}u}{u(u^2 - 1)^{1/2}}
$$
$$
+ \frac{c_0^2}{4\pi} \frac{H(x)y}{s^3} \int_{u=1}^{\infty} \frac{\exp(-su|y|/c_0)\mathrm{d}u}{(u^2 - 1)^{1/2}} \tag{G.21}
$$

To cast the integrals to the form resembling the Laplace transform, we next substitute

$$
u = c_0 \tau / |y| \tag{G.22}
$$

for $\tau \geq |y|/c_0$. In this way, we may write

$$
\hat{J}^{AB}(x, y, s) = \frac{yx^2 H(x)}{4\pi s^2} \int_{\tau=|y|/c_0}^{\infty} \exp(-s\tau) \frac{\mathrm{d}\tau}{\tau(\tau^2 - y^2/c_0^2)^{1/2}}
$$
$$
- \frac{yc_0^2 H(x)}{4\pi s^4} \int_{\tau=|y|/c_0}^{\infty} \exp(-s\tau) \frac{\mathrm{d}\tau}{\tau(\tau^2 - y^2/c_0^2)^{1/2}}
$$
$$
+ \frac{yc_0^2 H(x)}{4\pi s^3} \int_{\tau=|y|/c_0}^{\infty} \exp(-s\tau) \frac{\mathrm{d}\tau}{(\tau^2 - y^2/c_0^2)^{1/2}} \tag{G.23}
$$

which can be substituted in Eq. (G.19) to complete the expression for $\hat{J}^A(x, y, s)$. In the last step, Eq. (G.19) is transformed back to the TD. The transformation of Eq. (G.23) back to the TD is lengthy, yet straightforward, and the result can be written as

$$
J^{AB}(x, y, t) = \frac{\mathrm{sgn}(y)H(x)}{4\pi} \left\{ |y| \left(c_0^2 t^2 - x^2 + \frac{y^2}{3}\right) \cosh^{-1}\left(\frac{c_0 t}{|y|}\right) \right.
$$
$$
- c_0 t \left(\frac{c_0^2 t^2}{6} - x^2\right) \tan^{-1}\left[\left(\frac{c_0^2 t^2}{y^2} - 1\right)^{1/2}\right]
$$
$$
\left. - \frac{7}{6} c_0 t y^2 \left(\frac{c_0^2 t^2}{y^2} - 1\right)^{1/2} \right\} H(c_0 t - |y|) \tag{G.24}
$$

Finally, making use of Lerch's uniqueness theorem of the one-sided Laplace transformation [4, appendix], again, the TD counterpart of Eq. (G.19) is written as

$$
J^A(x, y, t) = \frac{1}{12\pi} \int_{v=r}^{c_0 t} (c_0 t - v)^3 \overline{F}(x, y, v)\mathrm{d}v + J^{AB}(x, y, t) \tag{G.25}
$$

where we defined $\overline{F}(x, y, v)$ from Eq. (G.20) via $v = c_0\tau$. Finally observe that owing to the elementary functional dependence of $\overline{F}(x, y, c_0\tau)$ on $c_0\tau$, the integration in Eq. (G.25) can also be carried out analytically (see Eq. (14.47)). As its integrand, however, does not show a singularity in $\tau \in [r/c_0, t]$ (recall that $x \neq 0$ and $y \neq 0$), we may evaluate this integral with the aid of the recursive convolution method (see appendix H) or using a standard integration routine.

G.1.2 Generic Integral J^B

We next transform Eq. (G.7) to the TD. To this end, the integrand is again continued analytically away from the imaginary axis, and the integration path $\mathbb{K}_0$ is replaced by the loop encircling the branch cut (see Figure G.1b). The new contour is, in fact, a limit of $\mathcal{G} \cup \mathcal{G}^*$ as defined via Eq. (G.9) for $|y| \downarrow 0$ and can be hence parametrized via

$$\kappa(\tau) = -\tau/x \pm i0 \tag{G.26}$$

for all $\tau \geq |x|/c_0$, for the path just above and below the branch cut, respectively. Combining the contributions from the latter paths, we arrive at

$$\hat{J}^B(x, y, s) = - (c_0 x/2\pi s^4)\mathrm{H}(y)$$
$$\times \int_{\tau=|x|/c_0}^{\infty} \exp(-s\tau)(c_0^2\tau^2 - x^2)^{1/2}\tau^{-3}\mathrm{d}\tau + \hat{J}^{BB} \tag{G.27}$$

in which $\hat{J}^{BB}$ denotes the corresponding contribution originating from the triple pole singularity at $\kappa = 0$ for $x > 0$. Applying Cauchy's formula again, the pole contribution can be evaluated at once. This way yields

$$\hat{J}^{BB}(x, y, s) = \left(\frac{c_0^3}{4s^4} - \frac{c_0 x^2}{4s^2}\right)\mathrm{H}(x)\mathrm{H}(y) \tag{G.28}$$

whose TD counterpart immediately follows

$$J^{BB}(x, y, t) = \frac{c_0 t}{4}\left(\frac{c_0^2 t^2}{6} - x^2\right)\mathrm{H}(x)\mathrm{H}(y)\mathrm{H}(t) \tag{G.29}$$

The latter expression can be finally substituted in the TD original of Eq. (G.27) that reads

$$J^B(x, y, t) = \frac{\mathrm{sgn}(x)\mathrm{H}(y)}{4\pi}\left\{|x|\left(c_0^2 t^2 - \frac{x^2}{6}\right)\cosh^{-1}\left(\frac{c_0 t}{|x|}\right)\right.$$
$$- c_0 t\left(\frac{c_0^2 t^2}{6} - x^2\right)\tan^{-1}\left[\left(\frac{c_0^2 t^2}{x^2} - 1\right)^{1/2}\right]$$
$$\left. - \frac{5}{3}c_0 t x^2\left(\frac{c_0^2 t^2}{x^2} - 1\right)^{1/2}\right\}\mathrm{H}(c_0 t - |x|) + J^{BB}(x, y, t) \tag{G.30}$$

Equations (G.25) with (G.24) and (G.30) with (G.29) can be hence used to express $J(x, y, t) = J^A(x, y, t) + J^B(x, y, t)$, which is the basic constituent of the impeditivity array (see Eq. (G.1)). When evaluating Eq. (G.2) for $y = w/2 > 0$, we use $\text{sgn}(y) - \text{sgn}(-y) = 2$ (see Eqs. (G.25) with (G.24) and (G.20)) and $H(y) - H(-y) = \text{sgn}(y) = 1$ (see Eqs. (G.30) with (G.29)).

APPENDIX H

A RECURSIVE CONVOLUTION METHOD AND ITS IMPLEMENTATION

The computationally most expensive part in filling the impeditivity matrix is the time-convolution integral in Eq. (G.25) that has been in section M.3 roughly approximated via its discrete form as offered by the MATLAB$^{\circledR}$ function conv. A more accurate, yet computationally efficient method, for calculating the time convolution is based on a recursive scheme. A description of the idea behind the recursive algorithm is the main purpose of the present chapter.

H.1 CONVOLUTION-INTEGRAL REPRESENTATION

The time convolution of two causal signals is defined as (cf. Eq. (1.2))

$$[f *_t g](t) = \int_{\tau=0}^{t} f(t - \tau)g(\tau)\mathrm{d}\tau \tag{H.1}$$

for all $t \geq 0$. As the error of a quadrature rule is proportional to (the power of) the time step, its computational burden grows with the increasing domain of integration $\tau \in [0, t]$. In order to avoid the difficulty, we next provide a recursive scheme that requires approximating an integral extending over the constant time step only. To that end, we assume that the Laplace-transform of $f(t)$ is known and can be used to represent $f(t - \tau)$ in Eq. (H.1) via the Bromwich inversion integral [3, section B.1], that is

$$[f *_t g](t) = \int_{\tau=0}^{t} \left[\frac{1}{2\pi\mathrm{i}} \int_{s\in\mathcal{B}} \exp[s(t - \tau)]\hat{f}(s)\mathrm{d}s \right] g(\tau)\mathrm{d}\tau \tag{H.2}$$

where $\mathcal{B}$ is the Bromwich contour that is parallel to $\mathrm{Re}(s) = 0$ and lies in the region of analyticity of $\hat{f}(s)$ in $\mathrm{Re}(s) > s_0$ (see Figure 1.1). Assuming now that $\hat{f}(s) = o(1)$ as $|s| \to \infty$, the Bromwich contour can be (for all $t > \tau$) supplemented with a

Time-Domain Electromagnetic Reciprocity in Antenna Modeling, First Edition. Martin Štumpf.
© 2020 by The Institute of Electrical and Electronics Engineers, Inc. Published 2020 by John Wiley & Sons, Inc.

semicircle of infinite radius to "close the contour" in the left-half s-plane, thereby replacing $\mathcal{B}$, in virtue of Cauchy's theorem, with the new (closed) integration path $\mathcal{C}$. Consequently, standard contour-integration methods (e.g. the residue theorem) can be applied to evaluate the integral in the complex s-plane. Changing next the order of the integrations with respect to τ and s, we get an equivalent expression

$$[f *_t g](t) = \frac{1}{2\pi i} \int_{s \in \mathcal{C}} \hat{G}(s,t) \hat{f}(s) \mathrm{d}s \tag{H.3}$$

whose inner integral, that is

$$\hat{G}(s,t) = \int_{\tau=0}^{t} \exp[s(t-\tau)]g(\tau)\mathrm{d}\tau \tag{H.4}$$

has the form amenable to the recursive representation [54]. It is then straightforward to demonstrate that

$$\hat{G}(s,t_k) = \hat{G}(s,t_{k-1}) \exp(s\Delta t) + \int_{\tau=t_{k-1}}^{t_k} \exp[s(t_k-\tau)]g(\tau)\mathrm{d}\tau \tag{H.5}$$

along the discretized time axis $\{t_k = k\Delta t; \Delta t > 0, k = 1, 2, \ldots, M\}$ using $\hat{G}(s,t) = 0$ for $t \le 0$. Equation (H.5) thus makes it possible to evaluate the inner integral of Eq. (H.3), for a parameter $s \in \mathcal{C}$, at any time $t = t_k$ using its previous state at $t = t_{k-1}$ and an integral term with the constant domain of integration $\Delta t = t_k - t_{k-1}$. The integral can be subsequently approximated using an appropriate quadrature rule. In this respect, Simpson's 3/8 rule leads to

$$\int_{\tau=t_{k-1}}^{t_k} \exp[s(t_k-\tau)]g(\tau)\mathrm{d}\tau$$

$$\simeq (\Delta t/8)g(t_{k-1})\exp(s\Delta t) + (3\Delta t/8)g(t_{k-1}+\Delta t/3)\exp(2s\Delta t/3)$$
$$+ (3\Delta t/8)g(t_{k-1}+2\Delta t/3)\exp(s\Delta t/3) + (\Delta t/8)g(t_k) \tag{H.6}$$

while the two-point Gauss-Legendre quadrature yields

$$\int_{\tau=t_{k-1}}^{t_k} \exp[s(t_k-\tau)]g(\tau)\mathrm{d}\tau$$

$$\simeq (\Delta t/2)g[t_{k-1}+(1-1/\sqrt{3})\Delta t/2]\exp[s(1+1/\sqrt{3})\Delta t/2]$$
$$+ (\Delta t/2)g[t_{k-1}+(1+1/\sqrt{3})\Delta t/2]\exp[s(1-1/\sqrt{3})\Delta t/2] \tag{H.7}$$

for instance. The values of $\hat{G}(s,t)$ obtained in this fashion are finally used in an approximation of the contour integral in Eq. (H.3). An illustrative example of the methodology is given in the following section.

H.2 ILLUSTRATIVE EXAMPLE

The handling of the time-convolution integral described in the previous section will next be illustrated on an example. For this purpose, we take the convolution integral that appeared in appendix G in the definition of the relevant impeditivity matrix. Hence, referring to Eq. (G.25), we shall evaluate the following convolution integral

$$\frac{1}{6\pi} \int_{v=0}^{c_0 t} (c_0 t - v)^3 \overline{g}(v) dv \tag{H.8}$$

assuming that $\overline{g}(c_0 t) = 0$ for all $c_0 t \leq r$. Since $f(t) = (t^3/6)\mathrm{H}(t)$ can be represented by $\hat{f}(s) = 1/s^4$ in the Laplace-transform domain, we may, in line with Eq. (H.3), write

$$\frac{1}{2\pi^2 \mathrm{i}} \int_{s \in \mathcal{C}} \hat{G}(s, c_0 t) s^{-4} ds \tag{H.9}$$

in which (see Eq. (H.4))

$$\hat{G}(s, c_0 t) = \int_{v=0}^{c_0 t} \exp[s(c_0 t - v)]\overline{g}(v) dv \tag{H.10}$$

and $\mathcal{C}$ represents, in view of the (fourfold) pole singularity at $s = 0$, a circular contour enclosing the origin (see Figure H.1). Taking $\varrho > 0$ as its radius, the integration contour can be parametrized via

$$\mathcal{C} = \{s = \varrho \exp(\mathrm{i}\psi); \varrho > 0, 0 \leq \psi < 2\pi\} \tag{H.11}$$

which leads to

$$\frac{1}{2\pi^2 \varrho^3} \int_{\psi=0}^{2\pi} \hat{G}[\varrho \exp(\mathrm{i}\psi), c_0 t] \exp(-3\mathrm{i}\psi) d\psi \tag{H.12}$$

Finally, approximating the integration with respect to the parameter ψ via the trapezoidal rule, for example, we end up with

$$\frac{1}{K\pi\varrho^3} \sum_{n=0}^{K-1} \hat{G}[\varrho \exp(\mathrm{i}\psi_n), c_0 t] \exp(-3\mathrm{i}\psi_n) \tag{H.13}$$

with $\psi_n = 2\pi n/K$, where K denotes the number of discretization points around $\mathcal{C}$. Since $\hat{G}(s, c_0 t)$ can be readily evaluated with the aid of the recursive procedure described in the previous section, Eq. (H.13) represents the desired approximation of the convolution integral (H.8). A sample MATLAB® implementation of the entire procedure is given in the ensuing part.

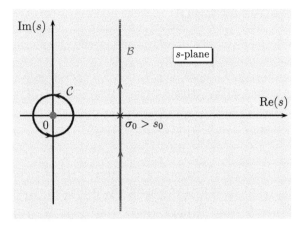

FIGURE H.1. Complex s-plane with the Bromwich contour $\mathcal{B}$ and the circular contour $\mathcal{C}$ around the origin.

H.3   IMPLEMENTATION OF THE RECURSIVE CONVOLUTION METHOD

The convolution integral from Eq. (G.25) has been in section M.3 approximated using the MATLAB® function conv, which is a relatively crude approximation. A more accurate way to calculate the integral employs the procedure described previously. Along these lines of reasoning, an alternative implementation of RF.m may be written in the following way:

```
1   % - - - - - - - - - - - - - - RF.m - - - - - - - - - - - - - - -
2   function out = RF(c0t,x,y)
3   %
4   M = length(c0t);
5   c0dt = c0t(2) - c0t(1);
6   %
7   rho = 1.0e-1; K = 6; psi = (0 : 2*pi/K: 2*pi-2*pi/K).';
8   s = rho*exp(1i*psi);
9   %
10  G = zeros(K, M);
11  k0 = find(c0t > sqrt(x^2+y^2), 1, 'first');
12  %
13  C = @(S)(@(V) F(x,y,V).*exp(S.*(c0t(k0)-V)));
14  CI = @(S) quadgk(C(S),c0t(k0-1), c0t(k0));
```

```
15   G(:, k0) = arrayfun(CI, s);
16   %
17   h = c0dt/4;
18   beta0 = 14*h/45*exp(s*c0dt);
19   beta1 = 64*h/45*exp(s*(c0dt-h));
20   beta2 = 24*h/45*exp(s*(c0dt-2*h));
21   beta3 = 64*h/45*exp(s*(c0dt-3*h));
22   beta4 = 14*h/45;
23   %
24   for k = k0+1 : M
25       %
26       G(:, k) = G(:, k-1).*exp(s*c0dt) ...
27           + beta0*F(x,y,c0t(k-1)) ...
28           + beta1*F(x,y,c0t(k-1)+h) ...
29           + beta2*F(x,y,c0t(k-1)+2*h) ...
30           + beta3*F(x,y,c0t(k-1)+3*h) ...
31           + beta4*F(x,y,c0t(k)));
32       %
33   end
34   %
35   G = G.*repmat(1./s.^3, [1 M]);
36   JA = real(sum(G))/K/pi;
37   %
38   JAB = (1/2/pi)*(y*(c0t.^2 - x^2 + y^2/3).*acosh(c0t/y) ...
39       - c0t.*(c0t.^2/6 - x^2).*atan(real(sqrt(c0t.^2/y^2 - 1))) ...
40       - (7/6)*y^2*c0t.*sqrt(c0t.^2/y^2 - 1))*(x > 0);
41   JAB = [zeros(1, M - length(JAB(c0t > y))) JAB(c0t > y)];
42   %
43   JB = (1/4/pi)*(abs(x)*(c0t.^2 - x^2/6).*acosh(c0t/abs(x)) ...
44       - c0t.*(c0t.^2/6 - x^2).*atan(real(sqrt(c0t.^2/x^2 - 1))) ...
45       - (5/3)*x^2*c0t.*sqrt(c0t.^2/x^2 - 1))*sign(x);
46   JB = [zeros(1, M - length(JB(c0t > abs(x)))) JB(c0t > abs(x))];
47   %
48   JBB = (c0t/4).*(c0t.^2/6 - x^2).*(x > 0).*(c0t > 0);
49   %
50   out = JA + JAB + JB + JBB;
```

Clearly, for $\varrho = 0.1$ s^{-1} and $K = 6$, we have first found the corresponding s-values lying on the circular contour $\mathcal{C}$. These values, that is, $s_n = \varrho \exp(2\mathrm{i}\pi n/K)$ for $n = \{0, 1, \ldots, 5\}$, are stored in variable s. Owing to the property that $F(x, y, \tau)$

changes rapidly close to $\tau = r/c_0 = (x^2 + y^2)^{1/2}/c_0$ (see Eq. (G.20)), the integration around the "arrival time" is carried out with the aid of the MATLAB$^{®}$ adaptive integration routine quadgk for all the chosen s_n values. In the ensuing for loop, we have implemented Eq. (H.5), in which the integration has been approximated via Boole's rule [25, eq. (25.4.14)]. Finally, the integration around the circular path C is approximated using the trapezoidal rule (see Eqs. (H.12) and (H.13)), and the result is stored in variable JA, again. The rest of the code remains the same as the one presented in section M.3.

APPENDIX I

CONDUCTANCE AND CAPACITANCE OF A THIN HIGH-CONTRAST LAYER

We shall analyze EM scattering from a narrow, infinitely long planar strip whose electric conductivity $\hat{\sigma} = \hat{\sigma}(z, s)$ and permittivity $\hat{\epsilon} = \hat{\epsilon}(z, s)$ show a high contrast with respect to the background medium described by $\{\epsilon_0, \mu_0\}$. The thickness of the strip is denoted by $\delta > 0$, and its width is $w > 0$ (see Figure I.1). Assuming that both the problem configuration as well as its EM excitation are x-independent, the EM field equations decouple into two independent sets from which we excite the transverse-magnetic (with respect to the strip axis) one only. Accordingly, we define the nonvanishing scattered EM fields as the difference between the total EM wave fields in the configuration and the incident EM wave fields, that is

$$\{\hat{E}_x^s, \hat{H}_y^s, \hat{H}_z^s\}(y, z, s) = \{\hat{E}_x, \hat{H}_y, \hat{H}_z\}(y, z, s)$$
$$- \{\hat{E}_x^i, \hat{H}_y^i, \hat{H}_z^i\}(y, z, s) \tag{I.1}$$

The cross-boundary conditions for the scattered EM fields are next derived based on the methodology introduced in Ref. [18]. In the domain occupied by the planar strip, the scattered field is governed by [3, section 18.3]

$$-\partial_y \hat{H}_z^s + \partial_z \hat{H}_y^s + s\epsilon_0 \hat{E}_x^s = -\hat{J}_x^c \tag{I.2}$$

$$\partial_z \hat{E}_x^s + s\mu_0 \hat{H}_y^s = 0 \tag{I.3}$$

$$-\partial_y \hat{E}_x^s + s\mu_0 \hat{H}_z^s = 0 \tag{I.4}$$

in which

$$\hat{J}_x^c = [s(\hat{\epsilon} - \epsilon_0) + \hat{\sigma}]\hat{E}_x \tag{I.5}$$

Time-Domain Electromagnetic Reciprocity in Antenna Modeling, First Edition. Martin Štumpf.
© 2020 by The Institute of Electrical and Electronics Engineers, Inc. Published 2020 by John Wiley & Sons, Inc.

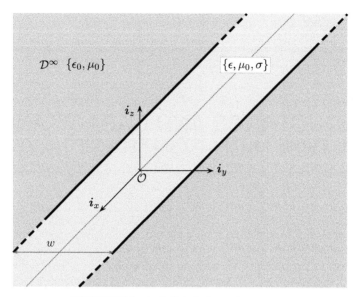

FIGURE I.1.   An infinitely-long planar strip.

is the equivalent electric-current contrast volume density. Carrying now the integrations of Eqs. (I.2) and (I.3) with respect to z, we find the desired cross-boundary conditions for the scattered field (cf. [18, eqs. (12)–(14)])

$$\lim_{z\downarrow\delta/2}\hat{E}_x^s(y,z,s)-\lim_{z\uparrow-\delta/2}\hat{E}_x^s(y,z,s)=O(\delta) \qquad (\text{I.6})$$

for all $y\in\mathbb{R}$ as $\delta\downarrow 0$ and

$$\lim_{z\downarrow\delta/2}\hat{H}_y^s(y,z,s)-\lim_{z\uparrow-\delta/2}\hat{H}_y^s(y,z,s)$$
$$=-[\hat{G}^L(s)+s\hat{C}^L(s)]\hat{E}_x(y,0,s)\Pi(y)+O(\delta) \qquad (\text{I.7})$$

for all $y\in\mathbb{R}$ as $\delta\downarrow 0$, $\Pi(y)$ was defined by Eq. (14.16), and

$$\hat{G}^L(s)=\int_{z=-\delta/2}^{\delta/2}\hat{\sigma}(z,s)\mathrm{d}z \qquad (\text{I.8})$$

$$\hat{C}^L(s)=\int_{z=-\delta/2}^{\delta/2}\hat{\epsilon}(z,s)\mathrm{d}z \qquad (\text{I.9})$$

are the conductance and capacitance parameters of the strip, respectively, that are $O(1)$ as $\delta\downarrow 0$. The axial component of the equivalent electric-current surface density is proportional to the jump of the y-component of the magnetic-field strength (see Eq. (I.7))

$$\partial \hat{J}^s_x(y, s) = - \lim_{z \downarrow \delta/2} \hat{H}^s_y(y, z, s) + \lim_{z \uparrow -\delta/2} \hat{H}^s_y(y, z, s) \qquad \text{(I.10)}$$

as $\delta \downarrow 0$. Finally, combination of Eqs. (I.7) and (I.10) yields the relation between the induced electric-current surface density and the total electric-field strength on the strip, that is

$$\partial \hat{J}^s_x(y, s) = [\hat{G}^L(s) + s\hat{C}^L(s)]\hat{E}_x(y, 0, s)\Pi(y) + O(\delta) \qquad \text{(I.11)}$$

for all $y \in \mathbb{R}$ as $\delta \downarrow 0$. This relation is used in chapter 15 to evaluate the impact of a finite permittivity and conductivity on EM scattering from a strip antenna.

APPENDIX J

GROUND-PLANE IMPEDITIVITY MATRIX

The elements of the TD impeditivity matrix accounting for the presence of the PEC ground plane can be found from (cf. Eq. (G.1))

$$
\begin{aligned}
\mathcal{K}^{[S,n]}(t) = [\zeta_0/c_0 \Delta t \Delta] \, [& P(x_S - x_n + 3\Delta/2, w/2, 2z_0, t) \\
& - 3P(x_S - x_n + \Delta/2, w/2, 2z_0, t) + 3P(x_S - x_n - \Delta/2, w/2, 2z_0, t) \\
& - P(x_S - x_n - 3\Delta/2, w/2, 2z_0, t)]
\end{aligned}
\tag{J.1}
$$

for all $S = \{1, \ldots, N\}$ and $n = \{1, \ldots, N\}$, where

$$
P(x, y, z, t) = I(x, y, z, t) - I(x, -y, z, t)
\tag{J.2}
$$

and $I(x, y, z, t)$ is determined with the aid of the CdH method [8] applied to the generic integral representation that is closely analyzed in the ensuing section.

J.1 GENERIC INTEGRAL I

The complex FD counterpart of $I(x, y, z, t)$ can be represented via

$$
\begin{aligned}
\hat{I}(x, y, z, s) = \frac{c_0^2}{8\pi^2} & \int_{\kappa \in \mathbb{K}_0} \frac{\exp(s\kappa x)}{s^3 \kappa^3} \Omega_0^2(\kappa) d\kappa \\
& \times \int_{\sigma \in \mathbb{S}_0} \frac{\exp\{-s[-\sigma y + \Gamma_0(\kappa, \sigma)z]\}}{s\sigma} \frac{d\sigma}{\Gamma_0(\kappa, \sigma)}
\end{aligned}
\tag{J.3}
$$

for $\{x \in \mathbb{R}; x \neq 0\}$, $\{y \in \mathbb{R}; y \neq 0\}$, $\{z \in \mathbb{R}; z \geq 0\}$, and $\{s \in \mathbb{R}; s > 0\}$. In Eq. (J.3), $\mathbb{K}_0$ and $\mathbb{S}_0$ denote, again, the indented integration path running along

Time-Domain Electromagnetic Reciprocity in Antenna Modeling, First Edition. Martin Štumpf.
© 2020 by The Institute of Electrical and Electronics Engineers, Inc. Published 2020 by John Wiley & Sons, Inc.

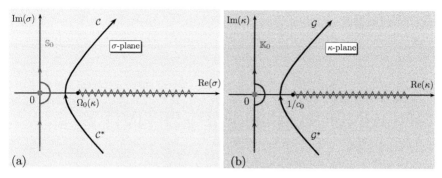

FIGURE J.1. (a) Complex σ-plane and (b) complex κ-plane with the original integration contours $\mathbb{S}_0$ and $\mathbb{K}_0$ and the new CdH paths for $y < 0$ and $x < 0$, respectively.

the imaginary axes in the complex κ- and σ-planes, respectively (see Figure J.1). Apparently, $\hat{J}(x, y, s)$ as given by Eq. (G.3) is a special case of $\hat{I}(x, y, z, s)$ for $z = 0$.

We next follow the CdH procedure similar to the ones described in sections F.1 and G.1. Hence, the integrand in the integral with respect to σ is first continued analytically into the complex σ-plane, while keeping $\text{Re}[\Gamma_0(\kappa, \sigma)] \geq 0$. This implies the branch cuts along $\{\text{Im}(\sigma) = 0, \Omega_0(\kappa) < |\text{Re}(\sigma)| < \infty\}$. Subsequently, making use of Jordan's lemma and Cauchy's theorem, the intended integration path $\mathbb{S}_0$ is deformed to the new CdH contour $\mathcal{C} \cup \mathcal{C}^*$ (see Figure J.1a) that is defined via (cf. Eq. (F.3))

$$-\sigma y + \Gamma_0(\kappa, \sigma)z = ud\,\Omega_0(\kappa) \tag{J.4}$$

for $\{1 \leq u < \infty\}$, and we defined $d \triangleq (y^2 + z^2)^{1/2} > 0$. Upon solving Eq. (J.4) for the slowness parameter in the y-direction, we get

$$\mathcal{C} = \left\{ \sigma(u) = \left[-yu/d + iz(u^2 - 1)^{1/2}/d \right] \Omega_0(\kappa) \right\} \tag{J.5}$$

for all $\{1 \leq u < \infty\}$. For $u = 1$, the contour intersects $\text{Im}(\sigma) = 0$ at $\sigma_0 = -(y/d)\Omega_0(\kappa)$, thus implying that $|\sigma_0| \leq \Omega_0(\kappa)$. Along the contour $\mathcal{C}$, we have

$$\Gamma_0[\kappa, \sigma(u)] = \left[zu/d + iy(u^2 - 1)^{1/2}/d \right] \Omega_0(\kappa) \tag{J.6}$$

and

$$\partial\sigma/\partial u = i\Gamma_0[\kappa, \sigma(u)]/(u^2 - 1)^{1/2} \tag{J.7}$$

for all $\{1 \leq u < \infty\}$. Subsequently, Schwartz's reflection principle is employed to combine the integrations along $\mathcal{C}$ and $\mathcal{C}^*$. Including further the pole contribution for $y > 0$, the inner integral of Eq. (J.3) can be written as

$$\frac{1}{2i\pi}\int_{\sigma\in\mathbb{S}_0}\frac{\exp\{-s[-\sigma y+\Gamma_0(\kappa,\sigma)z]\}}{s\sigma}\frac{d\sigma}{\Gamma_0(\kappa,\sigma)}=\frac{\exp[-s\Omega_0(\kappa)z]}{s\Omega_0(\kappa)}\mathrm{H}(y)$$

$$-\frac{1}{s\Omega_0(\kappa)}\frac{1}{\pi}\int_{u=1}^{\infty}\exp[-sud\Omega_0(\kappa)]\frac{yud}{u^2d^2-z^2}\frac{du}{(u^2-1)^{1/2}} \qquad (\text{J.8})$$

Equation (J.8) is subsequently substituted back in Eq. (J.3), which yields $\hat{I}=\hat{I}^A+\hat{I}^B$, where

$$\hat{I}^A(x,y,z,s)=\frac{1}{2\pi s}\int_{u=1}^{\infty}\frac{yud}{u^2d^2-z^2}\frac{du}{(u^2-1)^{1/2}}$$

$$\times\frac{c_0^2}{2i\pi}\int_{\kappa\in\mathbb{K}_0}\exp\{-s[-\kappa x+ud\Omega_0(\kappa)]\}\frac{\Omega_0(\kappa)}{s^3\kappa^3}d\kappa \qquad (\text{J.9})$$

and

$$\hat{I}^B(x,y,z,s)=-\frac{c_0^2}{2s}\frac{\mathrm{H}(y)}{2i\pi}\int_{\kappa\in\mathbb{K}_0}\frac{\exp\{-s[-\kappa x+\Omega_0(\kappa)z]\}}{s^3\kappa^3}\Omega_0(\kappa)d\kappa \qquad (\text{J.10})$$

The latter integral expressions will be next transformed to the TD in separate subsections.

J.1.1   Generic Integral I^A

In order to transform Eq. (J.9) to the TD, the integrand in the inner integral with respect to κ is continued analytically away from the imaginary axis, while keeping $\mathrm{Re}[\Omega_0(\kappa)]\geq 0$. This yields the branch cuts along $\{\mathrm{Im}(\kappa)=0,c_0^{-1}<|\mathrm{Re}(\kappa)|<\infty\}$ (see Figure J.1b). Relying further on Jordan's lemma and Cauchy's theorem, the integration path $\mathbb{K}_0$ is deformed into the new CdH path, say $\mathcal{G}\cup\mathcal{G}^*$, that is defined via

$$-\kappa x+ud\,\Omega_0(\kappa)=\tau \qquad (\text{J.11})$$

for $\{\tau\in\mathbb{R};\tau>0\}$. Upon solving Eq. (J.11) for the slowness parameter in the x-direction, we obtain

$$\mathcal{G}=\left\{\kappa(\tau)=[-x/R^2(u)]\tau+i[ud/R^2(u)][\tau^2-R^2(u)/c_0^2]^{1/2}\right\} \qquad (\text{J.12})$$

for all $\tau\geq R(u)/c_0$, and we defined $R(u)\triangleq(x^2+u^2d^2)^{1/2}>0$. Clearly, the CdH path intersects $\mathrm{Im}(\kappa)=0$ at $\kappa_0=-[x/R(u)]/c_0$, from which we get $|\kappa_0|<1/c_0$ (see Figure J.1b). Along $\mathcal{G}$, we may write

$$\Omega_0[\kappa(\tau)]=[ud/R^2(u)]\tau+i[x/R^2(u)][\tau^2-R^2(u)/c_0^2]^{1/2} \qquad (\text{J.13})$$

and

$$\partial\kappa/\partial\tau=i\Omega_0[\kappa(\tau)]/[\tau^2-R^2(u)/c_0^2]^{1/2} \qquad (\text{J.14})$$

for all $\tau\geq R(u)/c_0$. Combining the contributions from $\mathcal{G}$ and $\mathcal{G}^*$ and the pole singularity at $\kappa=0$, we end up with

$$\hat{I}^A(x,y,z,s) = \frac{c_0^3}{2\pi^2 s^4} \int_{u=1}^{\infty} \frac{yud}{z^2 - u^2 d^2} \frac{du}{(u^2-1)^{1/2}} \int_{\tau=R(u)/c_0}^{\infty} \exp(-s\tau)$$

$$\times \operatorname{Re}\left\{\frac{c_0^2\Omega_0^2[\kappa(\tau)]}{c_0^3\kappa^3(\tau)}\right\} \frac{d\tau}{[\tau^2 - R^2(u)/c_0^2]^{1/2}} + \hat{I}^{AB} \qquad \text{(J.15)}$$

where $\hat{I}^{AB}$ represents the pole contribution that is nonzero for $x > 0$. Next, using Eqs. (J.12) and (J.13) to express the integrand in the integral with respect to τ explicitly and interchanging the order of integration, we arrive at (cf. Eq. (G.13))

$$\hat{I}^A(x,y,z,s) = \frac{1}{2\pi^2} \frac{c_0^4}{s^4} \frac{1}{y} \int_{\tau=R/c_0}^{\infty} \exp(-s\tau)d\tau \int_{u=1}^{Q(\tau)} \frac{1}{u^2(d^2/y^2) - z^2/y^2}$$

$$\times \left[\frac{xc_0\tau}{c_0^2\tau^2 - u^2 d^2} - \frac{x^3 c_0^3 \tau^3}{(c_0^2\tau^2 - u^2 d^2)^3} - \frac{3x^3 c_0\tau u^2 d^2}{(c_0^2\tau^2 - u^2 d^2)^3}\right.$$

$$\left.+ \frac{3xc_0\tau u^2 d^2}{(c_0^2\tau^2 - u^2 d^2)^2}\right] \frac{u\,du}{[Q^2(\tau) - u^2]^{1/2}(u^2-1)^{1/2}} + \hat{I}^{AB} \quad \text{(J.16)}$$

where we defined $Q(\tau) = (c_0^2\tau^2 - x^2)^{1/2}/d > 0$ and $R = R(x,y,z) = R(1) = (x^2 + y^2 + z^2)^{1/2} > 0$. Again, substitution (A.9) is employed to carry out the inner integration with respect to u analytically. Transforming the inner integrals, we obtain

$$\frac{c_0\tau}{x} \int_{\psi=0}^{\pi/2} \frac{1}{\cos^2(\psi) + [(c_0^2\tau^2 - x^2 - z^2)/y^2]\sin^2(\psi)}$$

$$\times \frac{d\psi}{[(c_0^2\tau^2 - y^2 - z^2)/x^2]\cos^2(\psi) + \sin^2(\psi)}$$

$$- \frac{c_0^3\tau^3}{x^3} \int_{\psi=0}^{\pi/2} \frac{1}{\cos^2(\psi) + [(c_0^2\tau^2 - x^2 - z^2)/y^2]\sin^2(\psi)}$$

$$\times \frac{d\psi}{\{[(c_0^2\tau^2 - y^2 - z^2)/x^2]\cos^2(\psi) + \sin^2(\psi)\}^3}$$

$$- \frac{3d^2 c_0\tau}{x^3} \int_{\psi=0}^{\pi/2} \frac{\cos^2(\psi) + [(c_0^2\tau^2 - x^2)/(y^2 + z^2)]\sin^2(\psi)}{\cos^2(\psi) + [(c_0^2\tau^2 - x^2 - z^2)/y^2]\sin^2(\psi)}$$

$$\times \frac{d\psi}{\{[(c_0^2\tau^2 - y^2 - z^2)/x^2]\cos^2(\psi) + \sin^2(\psi)\}^3}$$

$$+ \frac{3d^2 c_0\tau}{x^3} \int_{\psi=0}^{\pi/2} \frac{\cos^2(\psi) + [(c_0^2\tau^2 - x^2)/(y^2 + z^2)]\sin^2(\psi)}{\cos^2(\psi) + [(c_0^2\tau^2 - x^2 - z^2)/y^2]\sin^2(\psi)}$$

$$\times \frac{d\psi}{\{[(c_0^2\tau^2 - y^2 - z^2)/x^2]\cos^2(\psi) + \sin^2(\psi)\}^2} \qquad \text{(J.17)}$$

to which we apply the following integral formulas

$$\int_{\psi=0}^{\pi/2} \frac{1}{\cos^2(\psi) + A^2\sin^2(\psi)} \frac{\mathrm{d}\psi}{B^2\cos^2(\psi) + \sin^2(\psi)}$$

$$= \frac{\pi}{2} \left[\frac{1}{A + A^2B} + \frac{1}{B + B^2A} \right] \tag{J.18}$$

with $A = (c_0^2\tau^2 - x^2 - z^2)^{1/2}/|y|$ and $B = (c_0^2\tau^2 - y^2 - z^2)^{1/2}/|x|$. Furthermore, we use

$$\int_{\psi=0}^{\pi/2} \frac{1}{\cos^2(\psi) + A^2\sin^2(\psi)} \frac{\mathrm{d}\psi}{[B^2\cos^2(\psi) + \sin^2(\psi)]^3}$$

$$= \frac{\pi}{16} \frac{1}{(AB+1)^3} \frac{1}{AB^5} \left[3AB^2(AB^5 + AB^3 + 1) \right.$$

$$\left. +9AB(B^5 + B^3 + AB^2 + A) + 8B^2(A^3 + B^3) + 3A \right] \tag{J.19}$$

and

$$\int_{\psi=0}^{\pi/2} \frac{\cos^2(\psi) + C^2\sin^2(\psi)}{\cos^2(\psi) + A^2\sin^2(\psi)} \frac{\mathrm{d}\psi}{[B^2\cos^2(\psi) + \sin^2(\psi)]^3}$$

$$= \frac{\pi}{16} \frac{1}{(AB+1)^3} \frac{1}{AB^5} \left\{ 3AB[B^2C^2(AB^4 + 3B^3 + 2B + A) \right.$$

$$+ B^3 + 2AB^2 + 3A] + 2B^2[B^3C^2(A^2 + 4) + A(4A^2 + 1)]$$

$$\left. +AB^2(AB^3 + C^2) + 3A \right\} \tag{J.20}$$

and, finally,

$$\int_{\psi=0}^{\pi/2} \frac{\cos^2(\psi) + C^2\sin^2(\psi)}{\cos^2(\psi) + A^2\sin^2(\psi)} \frac{\mathrm{d}\psi}{[B^2\cos^2(\psi) + \sin^2(\psi)]^2}$$

$$= \frac{\pi}{4} \frac{1}{(AB+1)^2} \frac{1}{AB^3} \left[B^2C^2(AB^2 + 2B + A) \right.$$

$$\left. +A(B^2 + 2AB + 1) \right] \tag{J.21}$$

Employing then formulas (J.18)–(J.21) in Eqs. (J.16) and (J.17), we arrive at

$$\hat{I}^A(x, y, z, s) = \frac{1}{2\pi} \frac{c_0^4}{s^4} \int_{\tau=R/c_0}^{\infty} \exp(-s\tau)V(x, y, z, \tau)\mathrm{d}\tau + \hat{I}^{AB} \tag{J.22}$$

in which the integral has the form of Laplace transformation. The additional term $\hat{I}^{AB}$ is determined upon evaluating the contribution from the (triple) pole singularity at $\kappa = 0$ in Eq. (J.9). Hence, employing Cauchy's formula [30, section 2.41], we end up with

$$\hat{I}^{AB}(x, y, z, s) = \frac{c_0}{4\pi} \frac{H(x)}{s^4} \left(s^2 x^2 - c_0^2\right) \int_{u=1}^{\infty} \frac{yud}{u^2 d^2 - z^2} \frac{\exp(-sud/c_0)du}{(u^2 - 1)^{1/2}}$$

$$+ \frac{c_0^2}{4\pi} \frac{H(x)y}{s^3} \int_{u=1}^{\infty} \frac{u^2 d^2}{u^2 d^2 - z^2} \frac{\exp(-sud/c_0)du}{(u^2 - 1)^{1/2}} \tag{J.23}$$

Again, the integrals in Eq. (J.23) are next cast into the form of Laplace transformation. To this end, we substitute

$$u = c_0 \tau / d \tag{J.24}$$

for $\tau \geq d/c_0$ and get

$$\hat{I}^{AB}(x, y, z, s) = \frac{y x^2 c_0 H(x)}{4\pi s^2} \int_{\tau=d/c_0}^{\infty} \exp(-s\tau) \frac{c_0 \tau}{c_0^2 \tau^2 - z^2} \frac{d\tau}{(\tau^2 - d^2/c_0^2)^{1/2}}$$

$$- \frac{y c_0^3 H(x)}{4\pi s^4} \int_{\tau=d/c_0}^{\infty} \exp(-s\tau) \frac{c_0 \tau}{c_0^2 \tau^2 - z^2} \frac{d\tau}{(\tau^2 - d^2/c_0^2)^{1/2}}$$

$$+ \frac{y c_0^2 H(x)}{4\pi s^3} \int_{\tau=d/c_0}^{\infty} \exp(-s\tau) \frac{c_0^2 \tau^2}{c_0^2 \tau^2 - z^2} \frac{d\tau}{(\tau^2 - d^2/c_0^2)^{1/2}} \tag{J.25}$$

The transformation of Eq. (J.25) is straightforward, yet relatively lengthy. Carrying out the resulting integrals, we end up with (cf. Eq. (G.24))

$$I^{AB}(x, y, z, t) = \frac{y/d}{4\pi} \left\{ d \left(c_0^2 t^2 - x^2 + \frac{d^2}{3} + \frac{2z^2}{3} \right) \cosh^{-1} \left(\frac{c_0 t}{d} \right) \right.$$

$$- \frac{c_0 t (c_0^2 t^2/6 - x^2 + 3z^2/2)}{(1 - z^2/d^2)^{1/2}} \tan^{-1} \left[\frac{(c_0^2 t^2/d^2 - 1)^{1/2}}{(1 - z^2/d^2)^{1/2}} \right]$$

$$+ \frac{z(c_0^2 t^2 - x^2 + 2z^2/3)}{(1 - z^2/d^2)^{1/2}} \tan^{-1} \left[\frac{z}{c_0 t} \frac{(c_0^2 t^2/d^2 - 1)^{1/2}}{(1 - z^2/d^2)^{1/2}} \right]$$

$$\left. - \frac{7}{6} c_0 t d^2 \left(\frac{c_0^2 t^2}{d^2} - 1 \right)^{1/2} \right\} H(x) H(c_0 t - d) \tag{J.26}$$

Finally, I^{AB} is substituted in the TD counterpart of Eq. (J.22), which yields

$$I^A(x, y, z, t) = \frac{1}{12\pi} \int_{v=R}^{c_0 t} (c_0 t - v)^3 \overline{V}(x, y, z, v) dv + I^{AB} \tag{J.27}$$

where $\overline{V}(x, y, z, v)$ corresponds to $V(x, y, z, \tau)$ with $v = c_0 \tau$. The convolution integral in Eq. (J.27) will be calculated with the aid of the recursive-convolution method (see appendix H).

J.1.2   Generic Integral I^B

In order to transform Eq. (J.10) to TD, we shall follow the strategy similar to the one from the previous section. In this way, the intended integration path $\mathbb{K}_0$ is replaced with the CdH-path $\mathcal{G} \cup \mathcal{G}^*$, again, that is defined now by (cf. Eq. (J.11))

$$-\kappa x + z\,\Omega_0(\kappa) = \tau \tag{J.28}$$

for all $\{\tau \in \mathbb{R}; \tau > 0\}$. The CdH-contour parametrization can be found by solving Eq. (J.28) as

$$\mathcal{G} = \left\{ \kappa(\tau) = (-x/\varrho^2)\tau + \mathrm{i}(z/\varrho^2)(\tau^2 - \varrho^2/c_0^2)^{1/2} \right\} \tag{J.29}$$

for all $\tau \geq \varrho/c_0$, where we defined $\varrho \triangleq (x^2 + z^2)^{1/2} > 0$. Along the path in the upper half of the complex κ-plane, we have

$$\Omega_0[\kappa(\tau)] = (z/\varrho^2)\tau + \mathrm{i}(x/\varrho^2)(\tau^2 - \varrho^2/c_0^2)^{1/2} \tag{J.30}$$

with the corresponding Jacobian

$$\partial\kappa/\partial\tau = \mathrm{i}\Omega_0[\kappa(\tau)]/(\tau^2 - \varrho^2/c_0^2)^{1/2} \tag{J.31}$$

for all $\tau \geq \varrho/c_0$. Employing further Schwartz's reflection principle to combine the contributions along $\mathcal{G}$ and $\mathcal{G}^*$, we obtain

$$\hat{I}^B(x,y,z,s) = -\frac{c_0^3\,\mathrm{H}(y)}{2\pi s^4} \int_{\tau=\varrho/c_0}^{\infty} \exp(-s\tau)$$

$$\times \mathrm{Re}\left\{ \frac{c_0^2\Omega_0^2[\kappa(\tau)]}{c_0^3\kappa^3(\tau)} \right\} \frac{\mathrm{d}\tau}{(\tau^2 - \varrho^2/c_0^2)^{1/2}} + \hat{I}^{BB} \tag{J.32}$$

where we take the values from Eqs. (J.29) and (J.30), and $\hat{I}^{BB}$ accounts for the pole contribution at $\kappa = 0$. Owing to the contour indentation (see Figure J.1b), the latter is nonzero for $x > 0$. Expressing the integrand in Eq. (J.32) in an explicit form using Eqs. (J.29) and (J.30), we get

$$\hat{I}^B(x,y,z,s) = -\frac{xc_0^3\mathrm{H}(y)}{2\pi s^4} \int_{\tau=\varrho/c_0}^{\infty} \exp(-s\tau)\frac{c_0\tau}{c_0^2\tau^2 - z^2}\frac{\mathrm{d}\tau}{(\tau^2 - \varrho^2/c_0^2)^{1/2}}$$

$$+ \frac{x^3 c_0^3\mathrm{H}(y)}{2\pi s^4} \int_{\tau=\varrho/c_0}^{\infty} \exp(-s\tau)\frac{c_0^3\tau^3}{(c_0^2\tau^2 - z^2)^3}\frac{\mathrm{d}\tau}{(\tau^2 - \varrho^2/c_0^2)^{1/2}}$$

$$+ \frac{3x^3 z^2 c_0^3\mathrm{H}(y)}{2\pi s^4} \int_{\tau=\varrho/c_0}^{\infty} \exp(-s\tau)\frac{c_0\tau}{(c_0^2\tau^2 - z^2)^3}\frac{\mathrm{d}\tau}{(\tau^2 - \varrho^2/c_0^2)^{1/2}}$$

$$- \frac{3xz^2 c_0^3\mathrm{H}(y)}{2\pi s^4} \int_{\tau=\varrho/c_0}^{\infty} \exp(-s\tau)\frac{c_0\tau}{(c_0^2\tau^2 - z^2)^2}\frac{\mathrm{d}\tau}{(\tau^2 - \varrho^2/c_0^2)^{1/2}} + \hat{I}^{BB} \tag{J.33}$$

Transforming the latter expression to the TD and carrying out the resulting integrals analytically, we after some algebra end up with (cf. Eq. (J.26))

$$
\begin{aligned}
I^B(x,y,z,t) = \frac{x/\varrho}{4\pi} \Bigg\{ & \varrho \left(c_0^2 t^2 - \frac{x^2}{6} + \frac{3z^2}{2} \right) \cosh^{-1}\left(\frac{c_0 t}{\varrho} \right) \\
& - \frac{c_0 t (c_0^2 t^2/6 - x^2 + 3z^2/2)}{(1 - z^2/\varrho^2)^{1/2}} \tan^{-1}\left[\frac{(c_0^2 t^2/\varrho^2 - 1)^{1/2}}{(1 - z^2/\varrho^2)^{1/2}} \right] \\
& + \frac{z(c_0^2 t^2 - x^2 + 2z^2/3)}{(1 - z^2/\varrho^2)^{1/2}} \tan^{-1}\left[\frac{z}{c_0 t} \frac{(c_0^2 t^2/\varrho^2 - 1)^{1/2}}{(1 - z^2/\varrho^2)^{1/2}} \right] \\
& - \frac{5}{3} c_0 t \varrho^2 \left(\frac{c_0^2 t^2}{\varrho^2} - 1 \right)^{1/2} \Bigg\} H(y) H(c_0 t - \varrho) + I^{BB}
\end{aligned}
\tag{J.34}
$$

and the pole contribution follows from Cauchy's formula [30, section 2.41]

$$
I^{BB}(x,y,z,t) = \frac{c_0 t - z}{4} \left(\frac{c_0^2 t^2}{6} - \frac{5z c_0 t}{6} + \frac{2z^2}{3} - x^2 \right)
$$
$$
\times H(x) H(y) H(c_0 t - z)
\tag{J.35}
$$

thus completing the TD counterpart of Eq. (J.3), that is, $I = I^A + I^B$.

APPENDIX K

IMPLEMENTATION OF CDH-MOM FOR THIN-WIRE ANTENNAS

In this section, a demo MATLAB$^{®}$ implementation of the CdH-MoM concerning a thin-wire antenna is given. The code is divided into blocks, each line of which is supplemented with its sequence number. For ease of compiling, the first line of blocks contains the name of file where the corresponding block is situated.

K.1   SETTING SPACE-TIME INPUT PARAMETERS

In the first step, we shall define variables describing the EM properties of vacuum. The EM wave speed in vacuum is exactly $c_0 = 299\ 792\ 458$ m/s, and the value of magnetic permeability, $\mu_0 = 4\pi \cdot 10^{-7}$ H/m, is fixed by the choice of the system of SI units. The electric permittivity ϵ_0 and the EM wave impedance ζ_0 then immediately follow. The corresponding MATLAB$^{®}$ code is

```
1   % - - - - - - - - - - - - - main.m - - - - - - - - - - - - - - - -
2   c0 = 299792458;
3   mu0 = 4*pi*1e-7;
4   ep0 = 1/mu0/c0^2;
5   zeta0 = sqrt(mu0/ep0);
```

and the corresponding variables are summarized in Table K.1.

In the next step, we set the configurational parameters of the analyzed problem. This is done by setting the length ℓ and the radius a of the analyzed wire antenna. In the present demo code, we take $\ell = 0.10$ m and $a = 0.10$ mm. Therefore, we write

---

Time-Domain Electromagnetic Reciprocity in Antenna Modeling, First Edition. Martin Štumpf.
© 2020 by The Institute of Electrical and Electronics Engineers, Inc. Published 2020 by John Wiley & Sons, Inc.

TABLE K.1. EM Constants and the Corresponding MATLAB® Variables

Name	Type	Description
c0	[1x1] double	EM wave speed in vacuum c_0 (m/s)
mu0	[1x1] double	Permeability in vacuum μ_0 (H/m)
ep0	[1x1] double	Permittivity in vacuum ϵ_0 (F/m)
zeta0	[1x1] double	EM wave impedance in vacuum $\zeta_0(\Omega)$

```
6    % - - - - - - - - - - - - - main.m - - - - - - - - - - - - - - -
7    l = 0.10;
8    a = 1.0e-4;
```

With the given antenna length, we may generate the spatial grid along its axis (see Figure 2.2a). To that end, we shall define the number of discretization points, N, excluding the antenna end points. Consequently, the spatial discretization step and the spatial grid follow from $\Delta = \ell/(N+1)$ and $z_n = -\ell/2 + n\,\Delta$ for $n = \{1, \ldots, N\}$, respectively. Hence, for 10 segments along the antenna, we set $N = 9$ and write

```
9    % - - - - - - - - - - - - - main.m - - - - - - - - - - - - - - -
10   N = 9;
11   dZ = 1/(N+1);
12   z = -1/2+dZ:dZ:1/2-dZ;
```

In a similar way, we next define the temporal variables. The upper bound of the time window of observation is chosen to be related to a multiple of the wire length, that is, we take max $c_0 t = 20\,\ell$, for instance. The (scaled) time step $c_0 \Delta t$ is then chosen to be a fraction of the spatial step Δ. For a stable output, we may take $c_0 \Delta t = \Delta/70$, for example, and write

```
13   % - - - - - - - - - - - - - main.m - - - - - - - - - - - - - - -
14   c0dt = dZ/70;
15   c0t = 0:c0dt:20*l;
16   M = length(c0t);
```

where M is the number of time points along the discretized time axis. For the reader's convenience, the corresponding MATLAB® variables are summarized in Table K.2.

**TABLE K.2. Spatial and Temporal Parameters and the Accompanying MATLAB®
Variables**

Name	Type	Description
l	[1x1] double	Length of antenna ℓ (m)
a	[1x1] double	Radius of antenna a (m)
N	[1x1] double	Number of inner nodes N (-)
z	[1xN] double	Positions of nodes z_n (m)
dZ	[1x1] double	Length of segments Δ (m)
M	[1x1] double	Number of time steps M (-)
c0t	[1xM] double	Scaled time axis $c_0 t$ (m)
c0dt	[1x1] double	Scaled time step $c_0 \Delta t$ (m)

K.2 ANTENNA EXCITATION

In this section, we will provide simple MATLAB® implementations of the antenna
excitation as given in section 2.4. The plane-wave and delta-gap types of excitation
are described separately.

K.2.1 Plane-Wave Excitation

With reference to Figure 2.1 and Eq. (2.17), the incident EM plane wave is specified by the plane-wave signature $e^i(t)$ and the polar angle of incidence θ. In the
present example, we take $\theta = 2\pi/5$ (rad), and as the excitation pulse, we choose a
truncated sine pulse shape with a unity amplitude. Accordingly, we write

```
17   % - - - - - - - - - - - - main.m - - - - - - - - - - - - - - - -
18   theta = 2*pi/5;
19   c0tw = 1.0*l;
20   q = 1.0;
21   ei = @(c0T) sin(2*pi*q*c0T/c0tw) .* ((c0T > 0).*(c0T<c0tw));
```

where q determines the number of periods in the interval bounded by the excitation pulse (scaled) time width c0tw. The latter is chosen to be equal to the antenna
length ℓ. Once the EM plane wave is defined, we may evaluate the excitation voltage array according to Eqs. (2.21) and (2.22). This can be done as written in the
following block

```
22    % - - - - - - - - - - - - - main.m  - - - - - - - - - - - - - -
23    V = zeros(N,M);
24    for s = 1 : N
25        %
26        zOFF(1) = 1/2 - z(s) - dZ;
27        zOFF(2) = 1/2 - z(s);
28        zOFF(3) = 1/2 - z(s) + dZ;
29        %
30        if (theta ~= pi/2)
31            %
32            V(s,:) = (-sin(theta)/dZ)*PW(ei,c0t,zOFF,theta);
33            %
34        else
35            %
36            V(s,:) = (-1/dZ)*PWp(ei,c0t,zOFF);
37            %
38        end
39        %
40    end
```

After initializing the excitation voltage array V, we evaluate each of its rows in a `for` loop. To this end, we call function `PW` if $\theta \neq \pi/2$ and `PWp` in the contrary case. These functions, directly corresponding to Eqs. (2.21) and (2.22), can be implemented as functions in separate files. For the former, we may write

```
1     % - - - - - - - - - - - - - PW.m - - - - - - - - - - - - - - - -
2     function out = PW(ei,c0t,zOFF,theta)
3     %
4     M = length(c0t);
5     c0dt = c0t(2) - c0t(1);
6     %
7     T = + (c0t - zOFF(1)*cos(theta)).*(c0t > zOFF(1)*cos(theta)) ...
8         - 2*(c0t - zOFF(2)*cos(theta)).*(c0t > zOFF(2)*cos(theta)) ...
9         + (c0t - zOFF(3)*cos(theta)).*(c0t > zOFF(3)*cos(theta));
10    %
11    C = conv(T,ei(c0t)); C = C(1:M);
12    %
13    out = c0dt*C/cos(theta)^2;
```

where we have simply approximated the time convolution from Eq. (2.21) using the MATLAB® function `conv`. Implementation of Eq. (2.22) applying to $\theta = \pi/2$ is even simpler and can be written as follows:

TABLE K.3. Plane-Wave Excitation and the Corresponding MATLAB$^{\circledR}$ Variabless

Name	Type	Description
ei	[1x1] function handle	Plane-wave signature $e^i(t)$ (V/m)
c0tw	[1x1] double	Scaled pulse time width $c_0 t_w$ (m)
theta	[1x1] double	Polar angle of incidence θ (rad)
V	[NxM] double	Excitation voltage array V (V)

```
1    % - - - - - - - - - - - - - - PWp.m - - - - - - - - - - - - - - - -
2    function out = PWp(ei,c0t,zOFF)
3    %
4    out = 0.5*ei(c0t)*(zOFF(1)^2 - 2*zOFF(2)^2 + zOFF(3)^2);
```

In this way, the excitation voltage array V can be filled. For the sake of convenience, a summary of the key variables of this subsection is given in Table K.3.

K.2.2 Delta-Gap Excitation

Referring again to Figure 2.1 and to Eq. (2.23), the delta-gap source is specified by the excitation voltage pulse shape $V^T(t)$ and by its position z_δ. In the present example, we place the source at the origin with $z_\delta = 0$, and the excitation voltage pulse is chosen to be a (bipolar) triangle of a unity amplitude, that is, we write

```
17   % - - - - - - - - - - - - - main.m - - - - - - - - - - - - - - -
18   zd = 0;
19   c0tw = 1.0*l;
20   VT = @(c0T) (2/c0tw)*(c0T.*(c0T>0) - 2*(c0T-c0tw/2).*(c0T>c0tw/2) ...
21       + 2*(c0T-3*c0tw/2).*(c0T>3*c0tw/2) - (c0T-2*c0tw).*(c0T>2*c0tw));
```

Like in the previous subsection, the next step starts by initializing the excitation voltage array V, which is subsequently filled in a for loop. Concerning the delta-gap source of vanishing width to which Eq. (2.26) applies, we write

```
22   % - - - - - - - - - - - - - main.m - - - - - - - - - - - - - - -
23   V = zeros(N,M);
24   for s = 1 : N
25       %
26       V(s,:) = -(VT(c0t)/dZ)*((zd + dZ - z(s)) ...
```

TABLE K.4. Delta-Gap Excitation and the Corresponding MATLAB® Variables

Name	Type	Description
VT	[1x1] function handle	Excitation voltage pulse $V^T(t)$ (V)
zd	[1x1] double	Position of the source gap z_δ (m)

```
27            *((zd + dZ - z(s)) > 0) ...
28            - 2*(zd - z(s))*((zd - z(s)) > 0) ...
29            + (zd - dZ - z(s))*((zd - dZ - z(s)) > 0));
30        %
31    end
```

A generalization of the code to the gap source of a finite width $\delta > 0$ is a straight-forward application of Eq. (2.25). Finally, the additional MATLAB® variables are given in Table K.4.

K.3 IMPEDANCE MATRIX

In this section, we make use of Eqs. (C.12) and (C.13) to fill the TD impedance array. For this purpose, we start by initializing a three-dimensional array Z. The initialization is followed by a nested for loop statement, where the impedance array elements are filled using Eq. (C.12). In such a way, we write

```
32    % - - - - - - - - - - - - - main.m - - - - - - - - - - - - - - - -
33    Z = zeros(N,N,M);
34    for s = 1 : N
35        for n = 1 : s
36            %
37            zOFF(1) = z(s) - z(n) + 2*dZ;
38            zOFF(2) = z(s) - z(n) + dZ;
39            zOFF(3) = z(s) - z(n);
40            zOFF(4) = z(s) - z(n) - dZ;
41            zOFF(5) = z(s) - z(n) - 2*dZ;
42            %
43            Z(s,n,:) = IF(c0t,zOFF(1),a) - 4*IF(c0t,zOFF(2),a) ...
44                      + 6*IF(c0t,zOFF(3),a) - 4*IF(c0t,zOFF(4),a) ...
```

```
45                      + IF(c0t,z0FF(5),a);
46              Z(n,s,:) = Z(s,n,:);
47              %
48          end
49      end
50      Z = (zeta0/2/pi/c0dt/dZ^2)*Z;
```

where the sequence numbers follow the numbering of the preceding subsection implementing the delta-gap source. The inner function IF directly corresponds to Eq. (C.13) and can be implemented along the following lines

```
1    % - - - - - - - - - - - - - - IF.m - - - - - - - - - - - - - -
2    function out = IF(c0t,z,r)
3    %
4    aZ = abs(z); D = c0t/aZ;
5    %
6    if (aZ ~= 0)
7        %
8        IBC = (aZ^3/12) ...
9            * (7/3 + log(D) - 3*D.^2.*(log(D) - 1) - 6*D + (2/3)*D.^3) ...
10           .* (c0t > aZ);
11       IBC(1) = 0;
12       %
13   else
14       %
15       IBC = (c0t.^3/18) .* (c0t > 0);
16       %
17   end
18   %
19   IP = z*(c0t.^2/2 - z^2/6).*acosh(c0t/r) ...
20       .* (c0t > r)*(z > 0) ...
21       - c0t.^2*z*(z > 0);
22   %
23   out = IBC + IP;
```

At first, we have defined two auxiliary variables aZ and D that correspond to $|z|$ and $c_0 t/|z|$, respectively. Subsequently, we have evaluated the branch-cut contribution IBC considering its limiting value for $|z| \downarrow 0$. The value stored in variable IP then corresponds to the pole contribution (see Figure C.1). In the final step, the two contributions are added together.

Finally recall that the presented code is written for illustrative purposes and is hence not optimized for speed. A more efficient code in this sense would remove the redundancy in (repetitive) calling the function IF for identical spatial offsets, for instance.

K.4 MARCHING-ON-IN-TIME SOLUTION PROCEDURE

With the voltage-excitation and impedance arrays at our disposal, the unknown electric-current space-time distribution can be evaluated via the updating step-by-step procedure described by Eq. (2.15). This way requires to calculate the inverse of the impedance matrix at $t = \Delta t$, that is, of $\underline{Z}_1$. For this purpose, we could use the MATLAB$^{\circledR}$ function inv. As its use, however, is not recommended anymore, we shall achieve the solution by solving two triangular systems using the LU factorization. Starting with the initialization of the induced electric-current array I, we may rewrite Eq. (2.15) as follows:

```
51   % - - - - - - - - - - - - - main.m  - - - - - - - - - - - - - -
52   I = zeros(N,M);
53   %
54   [LZ, UZ]  = lu(Z(:,:,2));
55   H = LZ\V(:,2); I(:,2) = UZ\H;
56   %
57   for m = 2 : M-1
58       %
59       SUM = zeros(N,1);
60       for k = 1 : m - 1
61           %
62           SUM = SUM   ...
63               + (Z(:,:,m-k+2) - 2*Z(:,:,m-k+1) + Z(:,:,m-k))*I(:,k);
64           %
65       end
66       %
67       H = LZ\(V(:,m+1) - SUM); I(:,m+1) = UZ\H;
68       %
69   end
```

Once the procedure is executed, the variable I contains the values of electric-current coefficients $i_k^{[n]}$ for all $n = \{1, \ldots, N\}$ and $k = \{1, \ldots, M\}$. For the sake of completeness, the key variables and their definition are given in Table K.5.

TABLE K.5. **Marching-on-in-Time Procedure and the Corresponding MATLAB$^{®}$ Variables**

Name	Type	Description
Z	[NxNxM] double	Impedance array $\underline{Z}(\Omega)$
I	[NxM] double	Induced electric-current array I (A)

K.5   CALCULATION OF FAR-FIELD TD RADIATION CHARACTERISTICS

With the electric-current distribution at our disposal, we may further calculate the TD far-field EM radiation characteristics according to the methodology given in section 6.3. Hence, we first sum the time-shifted electric-current pulses at the nodal points according to Eq. (6.22). Then, for the radiation characteristics observed at $\theta = \pi/8$, for example, we can write

```
70    % - - - - - - - - - - - - - main.m - - - - - - - - - - - - - - -
71    thetaRAD = pi/8;
72    %
73    REF = max(z*cos(thetaRAD));
74    PHI = zeros(N, M);
75    for n = 1 : N
76        %
77        shift = round((z(n)*cos(thetaRAD) - REF)/c0dt);
78        %
79        PHI(n, 1-shift:end) = I(n, 1:end+shift);
80        %
81    end
82    %
83    PHIsum = dZ*sum(PHI);
```

TABLE K.6. **Radiation Characteristics and the Corresponding MATLAB$^{®}$ Variables**

Name	Type	Description
thetaRAD	[1x1] double	Polar angle of observation θ (rad)
PHIsum	[1xM] double	Far-field TD potential function $\Phi_z^\infty(\theta, t)$ (A·m)
EthINF	[1xM-1] double	Far-field TD radiation characteristic $E_\theta^\infty(\theta, t)$ (V)

where the auxiliary variable REF ensures that the first element of PHIsum corresponds to the maximum time advance. The polar component of the electric-field EM radiation characteristic follows from Eq. (6.20), that is

```
84  % - - - - - - - - - - - - - main.m  - - - - - - - - - - - - - - -
85  EthINF = zeta0*diff(PHIsum)/c0dt*sin(thetaRAD);
```

where we have simply replaced the time differentiation with the time difference. Finally, the key MATLAB® variables of this section are given in Table K.6.

APPENDIX L

IMPLEMENTATION OF VED-INDUCED THÉVENIN'S VOLTAGES ON A TRANSMISSION LINE

In this part, we provide an illustrative demo MATLAB® implementation of the Thévenin-voltage response due to an impulsive VED source. Again, the following demo code is divided into blocks, each line of which is supplemented with its sequence number. For ease of compiling, the first line of blocks contains the name of file where the corresponding block is situated.

L.1 SETTING SPACE-TIME INPUT PARAMETERS

As in section K.1, we begin with the definition of EM constants applying to vacuum (see Table K.1). Subsequently, we set configurational parameters, namely, the length of the line L and its location in terms of x_1, x_2, y_0, and z_0 (see Figure 13.1). The position of the VED source is determined by its height $h > 0$ above the ground. Assuming, for instance, the transmission line of length $L = 100\,\text{mm}$ that is placed along $\{x_1 = -3L/4 < x < x_2 = L/4, y_0 = -L/10, z_0 = 3L/200\}$ above the PEC ground plane, we may write

```
1    % - - - - - - - - - - - - - main.m - - - - - - - - - - - - -
2    c0 = 299792458;
3    mu0 = 4*pi*1e-7;
4    ep0 = 1/mu0/c0^2;
5    %
```

Time-Domain Electromagnetic Reciprocity in Antenna Modeling, First Edition. Martin Štumpf.
© 2020 by The Institute of Electrical and Electronics Engineers, Inc. Published 2020 by John Wiley & Sons, Inc.

**TABLE L.1. Spatial and Temporal Parameters and the Accompanying MATLAB®
Variables**

Name	Type	Description
L	[1x1] double	Length of transmission line L (m)
x1	[1x1] double	Location of line along x-axis x_1 (m)
x2	[1x1] double	Location of line along x-axis x_2 (m)
y0	[1x1] double	Location of line along y-axis y_0 (m)
z0	[1x1] double	Height of line above ground plane z_0 (m)
M	[1x1] double	Number of time steps M (-)
c0t	[1xM] double	Scaled time axis $c_0 t$ (m)

```
6    L = 0.100;
7    x1 = -3*L/4; x2 = L/4;
8    y0 = -L/10; z0 = 3*L/200;
```

In the ensuing step, we define the time window of observation. This can be done as
follows

```
9    % - - - - - - - - - - - - - - - main.m - - - - - - - - - - - - -
10   M = 1.0e+4;
11   c0t = linspace(0, 10*L, M);
```

where we have taken 10^4 of time steps, and the upper bound of the time window
is related to the transmission-line length via max $c_0 t = 10\ L$. The corresponding
MATLAB® variables are summarized in Table L.1.

L.2   SETTING EXCITATION PARAMETERS

In the next part of the code, we shall set the parameters describing the excitation
VED source. Referring to Figure 13.1, its position is specified by its height h over
the ground plane. The source signature can be described by $j(t) = \ell\, i(t)$, where
$i(t)$ is the electric-current pulse (in A) and ℓ denotes here the length of the (short)
fundamental dipole. The latter parameter is assumed to be small with respect to the
spatial support of the excitation pulse. Accordingly, we may write

TABLE L.2. Excitation Parameters and the Accompanying MATLAB® Variables

Name	Type	Description
h	[1x1] double	Height of VED above ground plane h (m)
dj	[1x1] function handle	First derivative of source pulse $\partial_t j(t)$ (A·m/s)
c0tw	[1x1] double	Scaled pulse time width $c_0 t_w$ (m)
ell	[1x1] double	Length of VED source ℓ (m)

```
12    % - - - - - - - - - - - - - main.m - - - - - - - - - - - - -
13    h = 13*L/200;
14    %
15    c0tw = 1.0*L; ell = z0;
16    dj = @(c0T) ell*c0*(4/c0tw^2)*(c0T.*(c0T>0) ...
17                - 2*(c0T-c0tw/2).*(c0T>c0tw/2) ...
18                + 2*(c0T-3*c0tw/2).*(c0T>3*c0tw/2) ...
19                - (c0T-2*c0tw).*(c0T>2*c0tw));
```

where we have taken $h = 13L/200$, with $c_0 t_w = L$ and $\ell = z_0$. Furthermore, function dj specifies $\partial_t j(t)$ that appears in Eqs. (13.23) and (13.26). It is straightforward to verify that the function handle defines a bipolar triangular pulse shape that corresponds to the time derivative of a unit electric-current pulse whose voltage equivalent is given by Eq. (7.4) (see also Fig. 7.2a). The corresponding MATLAB® variables are summarized in Table L.2.

L.3 CALCULATING THÉVENIN'S VOLTAGES

Once the configurational and excitation parameters are defined, we may proceed with the evaluation of Eqs. (13.21), (13.22), (13.27), and (13.28) describing the VED-induced Thévenin voltages on a transmission line over the PEC ground. This step will be accomplished via a function whose input arguments are the configurational and excitation parameters and that returns the desired voltage responses. Accordingly, we may write

```
20    % - - - - - - - - - - - - - main.m - - - - - - - - - - - - -
21    [V1G V2G] = VRESP(dj, x1, x2, y0, z0, h, c0t);
```

where V1G and V2G are [1 x M] arrays representing $V_1^G(t)$ and $V_2^G(t)$, respectively. Function VRESP will be specified next in a separate file.

In the first step, we evaluate the contributions from the horizontal section of the transmission line. Recall that these contributions are in Eqs. (13.21), (13.22), (13.27), and (13.28) accounted for by $Q(x_1|x_2, y, z, t)$ function. Hence, we may write

```
1   % - - - - - - - - - - - - - - VRESP.m - - - - - - - - - - - - - - -
2   function [VG1,VG2] = VRESP(dj, x1, x2, y0, z0, h, c0t)
3   %
4   if (z0 < h)
5       %
6       F1 = -Q(x1, x2, y0, h-z0, c0t) + Q(x1, x2, y0, z0+h, c0t);
7       F2 = -Q(-x2, -x1, y0, h-z0, c0t) + Q(-x2, -x1, y0, z0+h, c0t);
8       %
9   elseif (z0 > h)
10      %
11      F1 = Q(x1, x2, y0, z0-h, c0t) + Q(x1, x2, y0, z0+h, c0t);
12      F2 = Q(-x2, -x1, y0, z0-h, c0t) + Q(-x2, -x1, y0, z0+h, c0t);
13      %
14  end
```

Next, we calculate the contributions from the vertical sections of the transmission line that are associated with function $\mathcal{V}(x, y, t)$ as defined by Eqs. (13.24) and (13.25). To that end, we write

```
15  % - - - - - - - - - - - - - VRESP.m - - - - - - - - - - - - - - -
16  L = x2 - x1;
17  %
18  H1 = V(x1, y0, z0, h, c0t) - V(x2, y0, z0, h, c0t - L);
19  H2 = V(x2, y0, z0, h, c0t) - V(x1, y0, z0, h, c0t - L);
```

where we have calculated the length of the transmission line L (see Table L.1). Finally, the contributions are added together, and the time convolutions with the (time derivative of the) source signature (see Eqs. (13.23) and (13.26)) are carried out. A MATLAB® code following this strategy can be written as

```
20  % - - - - - - - - - - - - - VRESP.m - - - - - - - - - - - - - - -
21  VG1 = F1 + H1; VG2 = F2 + H2;
22  %
```

```
23   c0dt = c0t(2) - c0t(1);
24   M = length(c0t);
25   %
26   VG1 = conv(dj(c0t), VG1)*c0dt/c0;
27   VG1 = VG1(1:M);
28   %
29   VG2 = conv(dj(c0t), VG2)*c0dt/c0;
30   VG2 = VG2(1:M);
```

where we used MATLAB$^{\circledR}$ function conv to approximate the continuous time-convolution operator. A more accurate implementation would involve built-in integration routines to calculate the time-convolution integrals.

It remains to specify inner functions Q and V that are applied inside VRESP function. Apart from the time convolutions that are calculated at the end of VRESP, these functions match with $\mathcal{Q}(x_1|x_2, y, z, t)$ and $\mathcal{V}(x, y, t)$ defined by Eqs. (13.23)–(13.25). Starting with the former one, we write

```
1    % - - - - - - - - - - - - - - - Q.m - - - - - - - - - - - - - - -
2    c0 = 299792458;
3    mu0 = 4*pi*1e-7;
4    zeta0 = mu0*c0;
5    %
6    L = x2 - x1;
7    %
8    out = zeta0*(I(x2, y, z, c0t - L) - I(x1, y, z, c0t));
9    %
10   function f = I(x,y,z,c0t)
11   f = z/4/pi/(y^2 + z^2)*P(x,y,z,c0t) ...
12       .*(c0t > sqrt(x^2+y^2+z^2));
```

where we have incorporated, for the sake of similarity with Eq. (13.23), a nested function called I that represents Eq. (F.19). The space-time function P(x,y,z,c0t) then clearly refers to Eq. (F.18) and can be defined in a new file, that is

```
1    % - - - - - - - - - - - - - - P.m - - - - - - - - - - - - - - -
2    function out = P(x, y, z, c0t)
3    %
```

```
4   d = sqrt(y^2 + z^2);
5   R = sqrt(d^2 + x^2);
6   %
7   out = (x*c0t - x^2 ...
8       - (R*(c0t.^2 - x^2) + c0t*d^2)./(R + c0t) ...
9       + c0t.^2*d^2/R^2)/R./c0t;
10  out(c0t == 0) = 0;
```

The next inner function from VRESP represents Eqs. (13.24) and (13.25), and its MATLAB® implementation may read

```
1   % - - - - - - - - - - - - - - V.m - - - - - - - - - - - - - - - -
2   function out = V(x, y, z0, h, c0t)
3   %
4   if (z0 < h)
5       %
6       out = U(x, y, h - z0, c0t) - U(x, y, z0 + h, c0t);
7       %
8   elseif (z0 > h)
9       %
10      out = 2*U(x, y, 1.0e-9, c0t) - U(x, y, z0 - h, c0t) ...
11          - U(x, y, z0 + h, c0t);
12      %
13  end
```

where $\mathcal{U}(x, y, 0, t)$ in Eq. (13.25) has been replaced with a limit $\lim_{\delta \downarrow 0} \mathcal{U}(x, y, \delta, t)$ to avoid the inverse square-root singularity due to Eq. (F.35). A more sophisticated approach to handle the issue is based on a stretching of the variable of integration [55, section VIII]. Finally, the space-time function U(x, y, z, c0t) is closely related to Eq. (13.26) and can be implemented in a new file, again. In this way, we may write

```
1   % - - - - - - - - - - - - - - U.m - - - - - - - - - - - - - - - -
2   function out = U(x, y, z, c0t)
3   %
4   c0 = 299792458;
5   mu0 = 4*pi*1e-7;
6   zeta0 = mu0*c0;
7   %
```

```
8    out = zeta0*J(x,y,z,c0t);
9    %
10   function f = J(x,y,z,c0t)
11   f = (1/4/pi)*(1./sqrt(c0t.^2 - x^2 - y^2) ...
12       - z*c0t/sqrt(x^2+y^2+z^2)^3) ...
13       .*(c0t > sqrt(x^2+y^2+z^2));
```

where we defined a nested function, again, to clearly show its link to $\mathcal{J}(x,y,z,t)$ as defined by Eq. (F.35).

L.4   INCORPORATING GROUND LOSSES

In case it is necessary to include ground losses, we may implement the approximate formulas (13.40) and (13.42) based on the Cooray-Rubinstein model. To that end, we may add the following code (see section L.3)

```
22   % - - - - - - - - - - - - - main.m - - - - - - - - - - - -
23   ep1 = 10.0*ep0; sigma = 1.0e-3;
24   %
25   d2j = @(c0T) ell*c0^2*(4/c0tw^2)*((c0T>0) ...
26              - 2*(c0T>c0tw/2) + 2*(c0T>3*c0tw/2) ...
27              - (c0T>2*c0tw));
28   %
29   [dV1G dV2G] = dVRESP(d2j, x1, x2, y0, z0, h, c0t, ep1, sigma);
```

where the meaning of the additional parameters is explained in Table L.3. In the latter piece of code, we used function dVRESP that returns variables called dV1G and dV2G. These arrays correspond to $\Delta V_1^G(t)$ and $\Delta V_2^G(t)$ from Eqs. (13.40) and (13.42), respectively. The function can be implemented as follows

```
1    % - - - - - - - - - - - - dVRESP.m - - - - - - - - - - - - -
2    function [dVG1,dVG2] = dVRESP(d2j, x1, x2, y0, z0, h, c0t, ep1, sigma)
3    %
4    c0 = 299792458;
5    mu0 = 4*pi*1e-7;
6    ep0 = 1/mu0/c0^2;
```

TABLE L.3. Additional Parameters and the Accompanying MATLAB® Variables

Name	Type	Description
ep1	[1x1] double	Electric permittivity of ground ϵ_1 (F/m)
sigma	[1x1] double	Electric conductivity of ground σ (S/m)
zeta1	[1x1] double	Wave impedance $\zeta_1 = (\mu_0/\epsilon_1)^{1/2} (\Omega)$
d2j	[1x1] function handle	Second derivative of source pulse $\partial_t^2 j(t) (\mathrm{A \cdot m/s^2})$

```
7    %
8    zeta1 = sqrt(mu0/ep1); L = x2 - x1;
9    %
10   dVG1 = K(x1, y0, z0+h, c0t) - K(x2, y0, z0+h, c0t - L);
11   dVG2 = K(-x2, y0, z0+h, c0t) - K(-x1, y0, z0+h, c0t - L);
```

where the variables dV1G and dV2G have been initially filled by the values matching the expressions between the square brackets in Eqs. (13.40) and (13.42). To that end, we used a nested function K that will be specified latter. It then remains to calculate the time convolution with $Z(t) *_t \partial_t j(t) = \partial_t^{-1} Z(t) *_t \partial_t^2 j(t)$, where $\partial_t^{-1} Z(t) = \zeta_1 I_0(\alpha t/2) \mathrm{H}(t)$ (see Eq. (13.41)). Approximating the continuous time-convolution operator via the built-in function conv, again, we may write

```
12   % - - - - - - - - - - - - - dVRESP.m - - - - - - - - - - - - - -
13   c0dt = c0t(2) - c0t(1);
14   M = length(c0t);
15   %
16   C = conv(besseli(0, sqrt(ep0/ep1)*sigma*zeta1*c0t/2, 1),d2j(c0t));
17   C = zeta1*C(1:M)*c0dt/c0;
```

where the auxiliary variable C contains $\zeta_1 \partial_t^{-1} Z(t) *_t \partial_t^2 j(t)$ and we used $\alpha t = (\epsilon_0/\epsilon_1)^{1/2} \sigma \zeta_1 c_0 t$ in the built-in MATLAB® function besseli. Next, we calculate the remaining time convolutions and define the nested function according to Eq. (F.43). This can be done along the following lines

```
18  % - - - - - - - - - - - - - dVRESP.m - - - - - - - - - - - - - -
19  dVG1 = conv(C, dVG1)*c0dt/c0;
20  dVG1 = dVG1(1:M);
21  %
22  dVG2 = conv(C, dVG2)*c0dt/c0;
23  dVG2 = dVG2(1:M);
24  %
25  function f = K(x,y,z,c0t)
26  f = 1/2/pi/sqrt(x^2+y^2+z^2)*H(x,y,z,c0t) ...
27      .*(c0t > sqrt(x^2+y^2+z^2));
```

In the ensuing step, we specify the space-time function $H(x,y,z,c0t)$ using Eq. (F.42) and write

```
1  % - - - - - - - - - - - - - - - H.m - - - - - - - - - - - - - - -
2  function out = H(x, y, z, c0t)
3  %
4  d = sqrt(y^2 + z^2);
5  R = sqrt(d^2 + x^2);
6  %
7  out = 1./(c0t + x).^2 .*(c0t.*(c0t + 2*x) - x*R ...
8      - (R*(c0t.^2 - x^2) + c0t*d^2)./(R + c0t));
```

The total voltage responses including the impact of the finite ground conductivity are finally found as

```
30  % - - - - - - - - - - - - - main.m - - - - - - - - - - - - - -
31  V1G = V1G + dV1G;
32  V2G = V2G + dV2G;
```

which completes the description of the demo MATLAB® implementation.

IMPLEMENTATION OF CDH-MOM FOR NARROW-STRIP ANTENNAS

In this section, a demo MATLAB® implementation of the CdH-MoM concerning a narrow-strip antenna is given. The code is divided into blocks, each line of which is supplemented with its sequence number. For ease of compiling, the first line of blocks contains the name of file where the corresponding block is situated.

M.1   SETTING SPACE-TIME INPUT PARAMETERS

As in section K.1, the MATLAB® code starts with the definition of EM constants, that is

```
1   % - - - - - - - - - - - - - - main.m - - - - - - - - - - - - -
2   c0 = 299792458;
3   mu0 = 4*pi*1e-7;
4   ep0 = 1/mu0/c0^2;
5   zeta0 = sqrt(mu0/ep0);
```

in accordance with Table K.1. In the ensuing section of the code, we define the length of the strip, denoted by ℓ, and its width w. In the demo code, we choose $\ell = 0.10\,\text{m}$ and $w = 5.0\,\text{mm}$, for instance. Therefore, we write

```
6   % - - - - - - - - - - - - - - main.m - - - - - - - - - - - - -
7   l = 0.10;
8   w = 5.0e-3;
```

Next, we define spatial grid points along the axis of the strip (see Figure 14.2). This can be done, for example, by defining the number of inner grid points, N,

---

Time-Domain Electromagnetic Reciprocity in Antenna Modeling, First Edition. Martin Štumpf.

through which we can calculate the spatial discretization step in the x-direction, $\Delta = \ell/(N + 1)$. The grid points along the axis then follow from $x_n = -\ell/2 + n \Delta$ for $n = \{1, \ldots, N\}$. The corresponding MATLAB® implementation with $N = 9$ can be written as follows

```
9   % - - - - - - - - - - - - - - main.m - - - - - - - - - - - - - - -
10  N = 9;
11  dX = 1/(N+1);
12  x = -1/2+dX:dX:1/2-dX;
```

Subsequently, we define the relevant temporal parameters. The upper bound of the time window of observation will be related to a multiple of the strip length. We take $\max c_0 t = 20 \ell$, for example. The (scaled) time step $c_0 \Delta t$ is again chosen to be a fraction of the spatial step Δ. In the present example, we take $c_0 \Delta t = \Delta/40$, for example, and write

```
13  % - - - - - - - - - - - - - - main.m - - - - - - - - - - - - - - -
14  c0dt = dX/40;
15  c0t = 0:c0dt:20*1;
16  M = length(c0t);
```

where M is the number of time points along the discretized time axis. The corresponding MATLAB® variables are given in Table M.1.

TABLE M.1. Spatial and Temporal Parameters and the Accompanying MATLAB® Variables

Name	Type	Description
l	[1x1] double	Length of antenna ℓ (m)
w	[1x1] double	Width of antenna w (m)
N	[1x1] double	Number of inner nodes N (-)
x	[1xN] double	Positions of nodes x_n (m)
dX	[1x1] double	Length of segments Δ (m)
M	[1x1] double	Number of time steps M (-)
c0t	[1xM] double	Scaled time axis $c_0 t$ (m)
c0dt	[1x1] double	Scaled time step $c_0 \Delta t$ (m)

M.2 DELTA-GAP ANTENNA EXCITATION

The incorporation of the delta-gap excitation has been described in section 14.4.2. In the present section, it is shown how this voltage source can be implemented in the demo code. As in section K.2.2, the delta-gap source is defined by its position x_δ along the strip's axis and its pulse shape $V^T(t)$ (see Figure 14.1). Taking $x_\delta = 0$ and choosing the (bipolar) triangular pulse shape with $c_0 t_w = \ell$ and the unit amplitude (cf. Eq. (9.3)), we may write

```
17    % - - - - - - - - - - - - - - main.m - - - - - - - - - - - -
18    xd = 0;
19    c0tw = 1.0*l;
20    VT = @(c0T) (2/c0tw)*(c0T.*(c0T>0) - 2*(c0T-c0tw/2).*(c0T>c0tw/2) ...
21        + 2*(c0T-3*c0tw/2).*(c0T>3*c0tw/2) - (c0T-2*c0tw).*(c0T>2*c0tw));
```

The successive part of the code starts by initializing the (space-time) excitation voltage $[N \times M]$ array, called V, whose elements will be filled using Eq. (14.30). A MATLAB® implementation of the latter can be then written as follows

```
22    % - - - - - - - - - - - - - main.m - - - - - - - - - - - - -
23    V = zeros(N,M);
24    for s = 1 : N
25        %
26        V(s,:) = -VT(c0t)*((xd + dX/2 - x(s) > 0) ...
27                         - (xd - dX/2 - x(s) > 0));
28        %
29    end
```

which applies to the voltage-gap source of vanishing width $\delta \downarrow 0$.

M.3 IMPEDITIVITY MATRIX

To calculate the space-time distribution of the sought electric-current surface density from Eq. (14.22), it is next necessary to calculate the elements of the TD impeditivity array. For this purpose, we shall rely on Eq. (G.1). The relevant MATLAB® code starts by initializing a three-dimensional $[N \times N \times M]$ space-time array, say Z. Its elements are subsequently filled in a nested loop statement, that is

```
30    % - - - - - - - - - - - - - main.m - - - - - - - - - - - - -
31    Z = zeros(N,N,M);
```

```
32   for s = 1 : N
33       for n = 1 : s
34           %
35               xOFF(1) = x(s) - x(n) + 3*dX/2;
36               xOFF(2) = x(s) - x(n) + dX/2;
37               xOFF(3) = x(s) - x(n) - dX/2;
38               xOFF(4) = x(s) - x(n) - 3*dX/2;
39           %
40               Z(s,n,:) = RF(c0t,xOFF(1),w/2) - 3*RF(c0t,xOFF(2),w/2) ...
41                        + 3*RF(c0t,xOFF(3),w/2) - RF(c0t,xOFF(4),w/2);
42               Z(n,s,:) = Z(s,n,:);
43           %
44       end
45   end
46   Z = (zeta0/c0dt/dX)*Z;
```

in which the function called RF directly corresponds to Eq. (G.2). A demo MATLAB® implementation of the latter is given on the following lines.

```
1    % - - - - - - - - - - - - - RF.m - - - - - - - - - - - - - - - -
2    function out = RF(c0t,x,y)
3    %
4    M = length(c0t);
5    c0dt = c0t(2) - c0t(1);
6    JA = (1/6/pi)*conv(c0t.^3,F(x,y,c0t))*c0dt;
7    JA = JA(1:M);
8    %
9    JAB = (1/2/pi)*(y*(c0t.^2 - x^2 + y^2/3).*acosh(c0t/y) ...
10        - c0t.*(c0t.^2/6 - x^2).*atan(real(sqrt(c0t.^2/y^2 - 1))) ...
11        - (7/6)*y^2*c0t.*sqrt(c0t.^2/y^2 - 1))*(x > 0);
12   JAB = [zeros(1, M - length(JAB(c0t > y))) JAB(c0t > y)];
13   %
14   JB = (1/4/pi)*(abs(x)*(c0t.^2 - x^2/6).*acosh(c0t/abs(x)) ...
15        - c0t.*(c0t.^2/6 - x^2).*atan(real(sqrt(c0t.^2/x^2 - 1))) ...
16        - (5/3)*x^2*c0t.*sqrt(c0t.^2/x^2 - 1))*sign(x);
17   JB = [zeros(1, M - length(JB(c0t > abs(x)))) JB(c0t > abs(x))];
18   %
19   JBB = (c0t/4).*(c0t.^2/6 - x^2).*(x > 0).*(c0t > 0);
20   %
21   out = JA + JAB + JB + JBB;
```

At first, we have initialized variables M and c0dt whose meaning is given in Table M.1. In the next step, we made use of a MATLAB$^{\circledR}$ function conv to approximate the time convolution integral that appears in Eq. (G.25). A more accurate, but computationally expensive, implementation may apply MATLAB$^{\circledR}$ built-in integration routines based on Gaussian quadratures. For example, we might replace lines 4 to 7 with the following code

```
4    for m = find(c0t > sqrt(x^2+y^2));
5        C = @(v) (c0t(m)-v).^3.*F(x,y,v);
6        JA(m) = (1/6/pi)*quadgk(C, sqrt(x^2+y^2), c0t(m));
7    end
```

that approximates the convolution integral at each time step using the Gauss-Kronrod quadrature. For a more efficient approach based on the recursive convolution method, we refer the reader to appendix H. Furthermore, due to the difference in Eq. (G.2) and the property $\overline{F}(x,y,v) = -\overline{F}(x,-y,v)$, the $[1 \times M]$ array JA is, in fact, filled by twice the values of the integration. The same also applies to array JAB whose values are found via Eq. (G.24). The function in the convolution integral, $\overline{F}(x,y,c_0t)$, is defined in a separate file. Referring to Eq. (G.20), we hence write

```
1    % - - - - - - - - - - - - - - F.m - - - - - - - - - - - - - - -
2    function out = F(x,y,c0t)
3    %
4    AX2 = c0t.^2/x^2; AY2 = c0t.^2/y^2; B2 = x^2/y^2;
5    %
6    out = 1/2./c0t.*(1./sqrt(AX2 - 1) + 1./sqrt(AY2 - 1)) ...
7        - y^4/16/x^2./c0t.^3./sqrt(AY2 - 1).^5.*(3*AY2.^4 ...
8        + 6*AY2.^3*(B2-1) + AY2.^2*(15*B2^2-10*B2+3) ...
9        + 4*AY2*B2*(1-5*B2) + 8*B2^2) ...
10       - x^2/2./c0t.^3./sqrt(AX2 - 1) ...
11       - 3*c0t/16/x^2./sqrt(AY2 - 1).^5.*(3*AY2.^2 ...
12       + 2*AY2*(B2-3) + B2*(3*B2-2) + 3) ...
13       + 3*c0t/4/x^2./sqrt(AY2 - 1).^3.*(AY2+B2-1);
14   %
15   out = out*sign(x);
16   out = out.*(c0t > sqrt(x^2+y^2)); out(isnan(out)) = 0;
```

where we have defined auxiliary variables AX2, AY2, and B2 corresponding to $c_0^2\tau^2/x^2$, $c_0^2\tau^2/y^2$, and x^2/y^2, respectively, to represent Eq. (G.20) in a concise form. Next, array JB apparently corresponds to the first term on the right-hand side of

TABLE M.2. Marching-on-in-Time Procedure and the Corresponding MATLAB®️ Variables

Name	Type	Description
Z	[NxNxM] double	Impeditivity array $\underline{\boldsymbol{Z}}(\Omega \cdot m)$
J	[NxM] double	Induced electric-current surface density array $\boldsymbol{J}$ (A/m)

Eq. (G.30), and JBB represents Eq. (G.29). Finally, the sum of the arrays on the last line of RF.m equals to $R(x, y, t)$ in agreement with Eq. (G.2).

M.4 MARCHING-ON-IN-TIME SOLUTION PROCEDURE

With the excitation and impeditivity arrays at out disposal, we may proceed with the evaluation of the electric-current surface density according to Eq. (14.22). The solution procedure that follows is similar to the one described in section K.4. Hence, along these lines, we start by initializing a $[N \times M]$ (space-time) array J for the coefficients of the induced electric-current surface density. Subsequently, employing the LU factorization to calculate the inverse, we may rewrite Eq. (14.22) as follows

```
47    % - - - - - - - - - - - - - main.m - - - - - - - - - - - -
48    J = zeros(N,M);
49    %
50    [LZ, UZ] = lu(Z(:,:,2));
51    H = LZ\V(:,2); J(:,2) = UZ\H;
52    %
53    for m = 2 : M-1
54        %
55        SUM = zeros(N,1);
56        for k = 1 : m - 1
57            %
58            SUM = SUM ...
59                + (Z(:,:,m-k+2) - 2*Z(:,:,m-k+1) + Z(:,:,m-k))*J(:,k);
60            %
61        end
62        %
63        H = LZ\(V(:,m+1) - SUM); J(:,m+1) = UZ\H;
64        %
65    end
```

Once the procedure is executed, the variable J contains the values of electric-current surface-density coefficients $j_k^{[n]}$ for all $n = \{1, \ldots, N\}$ and $k = \{1, \ldots, M\}$. In line with the expansion (14.15), the corresponding axial electric-current coefficients $i_k^{[n]}$ (cf. Eq. (2.12)) can simply be found via I $=$ w*J. Finally, the key variables of this section are given in Table M.2.

REFERENCES

1. A. T. de Hoop. Time-domain reciprocity theorems for electromagnetic fields in dispersive media. *Radio Science*, 22(7):1171–1178, 1987.
2. H. A. Lorentz. The theorem of Poynting concerning the energy in the electromagnetic field and two general propositions concerning the propagation of light. *Versl. Kon. Akad. Wetensch. Amsterdam*, 4:176, 1896.
3. A. T. de Hoop. *Handbook of Radiation and Scattering of Waves*. London, UK: Academic Press, 1995.
4. M. Štumpf. *Electromagnetic Reciprocity in Antenna Theory*. Hoboken, NJ: Wiley, 2018.
5. A. T. de Hoop. Reciprocity, discretization, and the numerical solution of direct and inverse electromagnetic radiation and scattering problems. *Proceedings IEEE*, 79(10):1421–1430, 1991.
6. G. Mur. Reciprocity and the finite-element modeling of electromagnetic wavefields. In P. M. Van den Berg, H. Blok, and J. T. Fokkema, editors, *Wavefields and Reciprocity (Proceedings of a Symposium held in honour of Professor Dr. A. T. de Hoop)*, pages 79–86, Delft, the Netherlands, November 1996.
7. M. Štumpf. *Pulsed EM Field Computation in Planar Circuits: The Contour Integral Method*. Boca Raton, FL: CRC Press, 2018.
8. A. T. de Hoop. A modification of Cagniard's method for solving seismic pulse problems. *Applied Scientific Research*, B(8):349–356, 1960.
9. A. T. de Hoop. Pulsed electromagnetic radiation from a line source in a two-media configuration. *Radio Science*, 14(2):253–268, 1979.

Time-Domain Electromagnetic Reciprocity in Antenna Modeling, First Edition. Martin Štumpf.
© 2020 by The Institute of Electrical and Electronics Engineers, Inc. Published 2020 by John Wiley & Sons, Inc.

10. A. T. de Hoop and J. H. M. T. Van der Hijden. Generation of acoustic waves by an impulsive point source in a fluid/solid configuration with a plane boundary. *The Journal of the Acoustical Society of America*, 75(6):1709–1715, 1984.

11. J. H. M. T. Van der Hijden. *Propagation of Transient Elastic Waves in Stratified Anisotropic Media*. PhD thesis, Delft University of Technology, the Netherlands, 1987.

12. A. T. de Hoop. Large-offset approximations in the modified Cagniard method for computing synthetic seismograms: A survey. *Geophysical Prospecting*, 36(5):465–477, 1988.

13. A. T. de Hoop. Transient two-dimensional Kirchhoff diffraction of a plane elastic SH wave by a generalized linear-slip fracture. *Geophysical Journal International*, 143(2):319–327, 2000.

14. J. R. Mosig. Integral equation technique. In T. Itoh, editor, Numerical Techniques for Microwave and Millimeter-Wave Passive Structures, *chapter 3*, pages 133–213. New York, NY: Wiley, 1989.

15. A. Sommerfeld. *Partial Differential Equations in Physics*. New York, NY: Academic Press, Inc., 1949.

16. M. Štumpf, A. T. de Hoop, and G. A. E. Vandenbosch. Generalized ray theory for time-domain electromagnetic fields in horizontally layered media. *IEEE Transactions on Antennas and Propagation*, 61(5):2676–2687, 2013.

17. C. Butler. The equivalent radius of a narrow conducting strip. *IEEE Transactions on Antennas and Propagation*, 30(4):755–758, 1982.

18. A. T. de Hoop and L. Jiang. Pulsed EM field response of a thin, high-contrast, finely layered structure with dielectric and conductive properties. *IEEE Transactions on Antennas and Propagation*, 57(8):2260–2269, 2009.

19. J. T. Fokkema and P. M. Van den Berg. *Seismic Applications of Acoustic Reciprocity*. Amsterdam: Elsevier, 1993.

20. D. V. Widder. *The Laplace Transform*. Princeton, NJ: Princeton University Press, 1946.

21. J. G. Van Bladel. *Electromagnetic Fields*. Hoboken, NJ: Wiley, 2nd edition, 2007.

22. R. W. P. King. The many faces of the insulated antenna. *Proceedings of the IEEE*, 64(2):228–238, 1976.

23. I. E. Lager, V. Voogt, and B. J. Kooij. Pulsed EM field, close-range signal transfer in layered configurations – A time-domain analysis. *IEEE Transactions on Antennas and Propagation*, 62(5):2642–2651, 2014.

24. G. D. Monteath. *Applications of the Electromagnetic Reciprocity Principle*. Oxford, UK: Pergamon Press, 1973.

25. M. Abramowitz and I. A. Stegun. *Handbook of Mathematical Functions*. New York, NY: Dover Publications, 1972.

26. M. Rubinstein. An approximate formula for the calculation of the horizontal electric field from lightning at close, intermediate, and long range. *IEEE Transactions on Electromagnetic Compatibility*, 38(3):531–535, 1996.

27. J. R. Wait. Concerning the horizontal electric field of lightning. *IEEE Transactions on Electromagnetic Compatibility*, 39(2):186, 1997.

28. A. T. de Hoop, I. E. Lager, and V. Tomassetti. The pulsed-field multiport antenna system reciprocity relation and its applications – A time-domain approach. *IEEE Transactions on Antennas and Propagation*, 57(3):594–605, 2009.

29. A. G. Tijhuis, P. Zhongqiu, and A. R. Bretones. Transient excitation of a straight thin-wire segment: A new look at an old problem. *IEEE Transactions on Antennas and Propagation*, 40(10):1132–1146, 1992.

30. E. C. Titchmarsh. *The Theory of Functions*. London, UK: Oxford University Press, 2nd edition, 1939.

31. F. M. Tesche. On the inclusion of loss in time-domain solutions of electromagnetic interaction problems. *IEEE Transactions on Electromagnetic Compatibility*, 32(1):1–4, 1990.

32. A. T. de Hoop and G. de Jong. Power reciprocity in antenna theory. *Proceedings of the Institution of Electrical Engineers*, 121(10):594–605, 1974.

33. M. Štumpf. Pulsed EM field radiation, mutual coupling and reciprocity of thin planar antennas. *IEEE Transactions on Antennas and Propagation*, 62(8):3943–3950, 2014.

34. A. K. Agrawal, H. J. Price, and S. H. Gurbaxani. Transient response of multiconductor transmission lines excited by a nonuniform electromagnetic field. *IEEE Transactions on Electromagnetic Compatibility*, 22(2):119–129, 1980.

35. F. M. Tesche, M. V. Ianoz, and T. Karlsson. *EMC Analysis Methods and Computational Models*. New York, NY: Wiley, 1997.

36. M. Štumpf. Pulsed vertical-electric-dipole excited voltages on transmission lines over a perfect ground – A closed-form analytical description. *IEEE Antennas and Wireless Propagation Letters*, 17(9):1656–1658, 2018.

37. F. Rachidi. A review of field-to-transmission line coupling models with special emphasis to lightning-induced voltages on overhead lines. *IEEE Transactions on Electromagnetic Compatibility*, 54(4):898–911, 2012.

38. F. Rachidi. Formulation of the field-to-transmission line coupling equations in terms of magnetic excitation field. *IEEE Transactions on Electromagnetic Compatibility*, 35(3):404–407, 1993.

39. C. Taylor, R. Satterwhite, and C. Harrison. The response of a terminated two-wire transmission line excited by a nonuniform electromagnetic field. *IEEE Transactions on Antennas and Propagation*, 13(6):987–989, 1965.

40. M. Štumpf and G. Antonini. Electromagnetic field coupling to a transmission line – A reciprocity-based approach. *IEEE Transactions on Electromagnetic Compatibility*, 2018.

41. D. M. Pozar. *Microwave Engineering*. New York, NY: Wiley, 2nd edition, 1997.

42. M. Leone and H. L. Singer. On the coupling of an external electromagnetic field to a printed circuit board trace. *IEEE Transactions on Electromagnetic Compatibility*, 41(4):418–424, 1999.

43. M. J. Master and M. A. Uman. Lightning induced voltages on power lines: Theory. *IEEE Transactions on Power Apparatus and Systems*, 103(9):2502–2518, 1984.

44. F. Rachidi, C. A. Nucci, M. V. Ianoz, and C. Mazzetti. Influence of a lossy ground on lightning-induced voltages on overhead lines. *IEEE Transactions on Electromagnetic Compatibility*, 38(3):250–264, 1996.

45. C. A. Nucci, F. Rachidi, M. V. Ianoz, and C. Mazzetti. Lightning-induced voltages on overhead lines. *IEEE Transactions on Electromagnetic Compatibility*, 35(1):75–86, 1993.

46. M. Štumpf and G. Antonini. Lightning-induced voltages on transmission lines over a lossy ground – An analytical coupling model based on the Cooray-Rubinstein formula. *IEEE Transactions on Electromagnetic Compatibility*, 2019.

47. M. Štumpf. Coupling of impulsive EM plane-wave fields to narrow conductive strips: An analysis based on the concept of external impedance. *IEEE Transactions on Electromagnetic Compatibility*, 60(2):548–551, 2018.

48. S. A. Schelkunoff. Modified Sommerfeld's integral and its applications. *Proceedings of the Institute of Radio Engineers*, 24(10):1388–1398, 1936.

49. S. A. Schelkunoff. *Electromagnetic Waves*. New York, NY: D. Van Nostrand Company, Inc., 9th printing edition, 1957.

50. K. C. Chen. Transient response of an infinite cylindrical antenna. *IEEE Transactions on Antennas and Propagation*, 31(1):170–172, 1983.

51. S. P. Morgan. Transient response of a dipole antenna. *Journal of Mathematical Physics*, 3(3):564–565, 1962.

52. T. T. Wu. Transient response of a dipole antenna. *Journal of Mathematical Physics*, 2(6):892–894, 1961.

53. M. Bingle, D. B. Davidson, and J. H. Cloete. Scattering and absorption by thin metal wires in rectangular waveguide – FDTD simulation and physical experiments. *IEEE Transactions on Microwave Theory and Techniques*, 50(6):1621–1627, 2002.

54. A. Semlyen and A. Dabuleanu. Fast and accurate switching transient calculations on transmission lines with ground return using recursive convolutions. *IEEE Transactions on Power Apparatus and Systems*, 94(2):561–571, 1975.

55. A. T. de Hoop, M. Štumpf, and I. E. Lager. Pulsed electromagnetic field radiation from a wide slot antenna with a dielectric layer. *IEEE Transactions on Antennas and Propagation*, 59(8):2789–2798, 2011.

INDEX

Time-Domain Electromagnetic Reciprocity in Antenna Modeling, First Edition. Martin Štumpf.
© 2020 by The Institute of Electrical and Electronics Engineers, Inc. Published 2020 by John Wiley & Sons, Inc.

IEEE PRESS SERIES ON ELECTROMAGNETIC WAVE THEORY